Calculus: An Introduction

Travis Madden

WILLFORD PRESS

www.willfordpress.com

Published by Willford Press,
118-35 Queens Blvd., Suite 400,
Forest Hills, NY 11375, USA

ISBN: 978-1-64728-347-6

Cataloging-in-Publication Data

Calculus : an introduction / Travis Madden.
    p. cm.
Includes bibliographical references and index.
ISBN 978-1-64728-347-6
1. Calculus. 2. Differential calculus. 3. Calculus, Integral.
4. Mathematics. 5. Statistics. I. Madden, Travis.
QA303.2 .C35 2022
515--dc23

For information on all Willford Press publications
visit our website at www.willfordpress.com

WILLFORD PRESS

TABLE OF CONTENTS

Permissions

Index

PREFACE

This book aims to help a broader range of students by exploring a wide variety of significant topics related to this discipline. It will help students in achieving a higher level of understanding of the subject and excel in their respective fields. This book would not have been possible without the unwavered support of my senior professors who took out the time to provide me feedback and help me with the process. I would also like to thank my family for their patience and support.

Calculus refers to the mathematical study of continuous change. The major branches of calculus are differential calculus and integral calculus. Differential calculus is concerned with the immediate rate of change and the slopes of curves. Integral calculus focuses on the accumulation of quantities and the areas under and between curves. Both branches are connected by the fundamental theorem of calculus. They utilize the fundamental concepts of convergence of infinite sequences and infinite series to a well-defined limit. Calculus is used in various branches of physical sciences, computer science, statistics, engineering, economics, business, actuarial science and demography. It is also used in various other fields where a problem is capable of being mathematically modeled and where an optimal solution is required. This textbook outlines the processes and applications of calculus in detail. It presents this complex subject in the most comprehensible and easy to understand language. The book will serve as a reference to a broad spectrum of readers.

A brief overview of the book contents is provided below:

Chapter – Calculus: An Introduction

The branch of mathematics which deals with the study of continuous change is known as calculus. It is broadly divided into two branches, differential calculus and integral calculus. Some of its key components are functions, their limits and their continuity. This is an introductory chapter which will introduce briefly all these significant aspects of calculus.

Chapter – Differential and Integral Calculus

Differential calculus is the branch of calculus which deals with the rate of change in quantities while integral calculus focuses on the accumulation of quantities and the areas under and between curves. Some of the types of integrals studied under this discipline are quadratic integrals, Borwein integrals and Dirichlet integrals. The diverse applications of differential and integral calculus have been thoroughly discussed in this chapter.

Chapter – Multivariable Calculus

The branch of calculus which deals with the differentiation and integration of functions involving several variables instead of one is known as multivariable calculus. A definite integral which is a function of more than one real variable is known as a multiple integral. This chapter discusses in detail these theories and methodologies related to multivariable calculus.

Chapter – Vector Calculus

The branch of calculus which deals with differentiation and integration of vector fields is known as vector calculus. Some of the differential operators studied within vector calculus are gradient, divergence and curl. All these diverse concepts related to vector calculus have been carefully analyzed in this chapter.

Chapter – Fundamental Theorems of Calculus

Some of the basic theorems in calculus are mean value theorem, Rolle's Theorem, extreme value theorem, Taylor's theorem and divergence theorem. Mean value theorem is used for proving statements regarding a function on an interval starting from local hypotheses about derivatives at points of the interval. This chapter has been carefully written to provide an easy understanding of the varied facets of these theories.

Travis Madden

Calculus: An Introduction

The branch of mathematics which deals with the study of continuous change is known as calculus. It is broadly divided into two branches, differential calculus and integral calculus. Some of its key components are functions, their limits and their continuity. This is an introductory chapter which will introduce briefly all these significant aspects of calculus.

Calculus is a branch of mathematics that involves the study of rates of change. Before calculus was invented, all math was static: It could only help calculate objects that were perfectly still. But the universe is constantly moving and changing. No objects—from the stars in space to subatomic particles or cells in the body—are always at rest. Indeed, just about everything in the universe is constantly moving. Calculus helped to determine how particles, stars, and matter actually move and change in real time.

Calculus is used in a multitude of fields that you wouldn't ordinarily think would make use of its concepts. Among them are physics, engineering, economics, statistics, and medicine. Calculus is also used in such disparate areas as space travel, as well as determining how medications interact with the body, and even how to build safer structures. You'll understand why calculus is useful in so many areas if you know a bit about its history as well as what it is designed to do and measure.

Calculus was developed in the latter half of the 17th century by two mathematicians, Gottfried Leibniz and Isaac Newton. Newton first developed calculus and applied it directly to the under-standing of physical systems. Independently, Leibniz developed the notations used in calculus. Put simply, while basic math uses operations such as plus, minus, times, and division (+, -, ×, and ÷), calculus uses operations that employ functions and integrals to calculate rates of change.

Those tools allowed Newton, Leibniz, and other mathematicians who followed to calculate things like the exact slope of a curve at any point. The Story of Mathematics explains the importance of Newton's fundamental theorem of the calculus.

Unlike the static geometry of the Greeks, calculus allowed mathematicians and engineers to make sense of the motion and dynamic change in the changing world around us, such as the orbits of planets, the motion of fluids, etc." Using calculus, scientists, astronomers, physicists, mathematicians, and chemists could now chart the orbit of the planets and stars, as well as the path of electrons and protons at the atomic level.

Branches of Calculus

There are two branches of calculus: differential and integral calculus. "Differential calculus studies the derivative and integral calculus studies the integral," the Massachusetts Institute of Technology.

But there is more to it than that. Differential calculus determines the rate of change of a quantity. It examines the rates of change of slopes and curves.

This branch is concerned with the study of the rate of change of functions with respect to their variables, especially through the use of derivatives and differentials. The derivative is the slope of a line on a graph. You find the slope of a line by calculating the rise over the run.

Integral calculus, by contrast, seeks to find the quantity where the rate of change is known. This branch focuses on such concepts as slopes of tangent lines and velocities. While differential calculus focuses on the curve itself, integral calculus concerns itself with the space or area under the curve. Integral calculus is used to figure the total size or value, such as lengths, areas, and volumes.

Calculus played an integral role in the development of navigation in the 17th and 18th centuries because it allowed sailors to use the position of the moon to accurately determine the local time. To chart their position at sea, navigators needed to be able to measure both time and angles with accuracy. Before the development of calculus, ship navigators and captains could do neither.

Calculus — both derivative and integral — helped to improve the understanding of this important concept in terms of the curve of the Earth, the distance ships had to travel around a curve to get to a specific location, and even the alignment of the Earth, seas, and ships in relation to the stars.

Practical Applications

Calculus has many practical applications in real life. Some of the concepts that use calculus include motion, electricity, heat, light, harmonics, acoustics, and astronomy. Calculus is used in geography, computer vision (such as for autonomous driving of cars), photography, artificial intelligence, robotics, video games, and even movies. Calculus is also used to calculate the rates of radioactive decay in chemistry, and even to predict birth and death rates, as well as in the study of gravity and planetary motion, fluid flow, ship design, geometric curves, and bridge engineering.

In physics, for example, calculus is used to help define, explain, and calculate motion, electricity, heat, light, harmonics, acoustics, astronomy, and dynamics. Einstein's theory of relativity relies on calculus, a field of mathematics that also helps economists predict how much profit a company or industry can make. And in shipbuilding, calculus has been used for many years to determine both the curve of the hull of the ship (using differential calculus), as well as the area under the hull (using integral calculus), and even in the general design of ships.

In addition, calculus is used to check answers for different mathematical disciplines such as statistics, analytical geometry, and algebra.

Calculus in Economics

Economists use calculus to predict supply, demand, and maximum potential profits. Supply and demand are, after all, essentially charted on a curve—and an ever-changing curve at that.

Economists use calculus to determine the price elasticity of demand. They refer to the ever-changing supply-and-demand curve as "elastic," and the actions of the curve as "elasticity." To calculate an exact measure of elasticity at a particular point on a supply or demand curve, you need to think about infinitesimally small changes in price and, as a result, incorporate mathematical derivatives into your elasticity formulas. Calculus allows you to determine specific points on that ever-changing supply-and-demand curve.

FUNCTION

A function from a set A to a set B is a rule that associates, to each element of A, a *unique* element of B.

Functions are typically denoted by lower-case or upper-case *single* letters, though some functions have special notations. To say that a function f is from A to B, we write $f : A \to B$ *is a function*.

For an element a of the domain A, the unique element of B associated with a is denoted as $f(a)$. The act of going from a to $f(a)$ is termed *applying* the function. The element a is termed an *input* to the function and the corresponding element $f(a)$ is termed the *output* or *image* of the function corresponding to that input.

A and B may be equal or distinct.

Some key terminology:

- The domain of a function $f : A \to B$ is the set A.

- This term is not used in most basic treatments of the calculus of one variable: The *co-domain* of a function $f : A \to B$ is the set B.

- The range of a function $f : A \to B$ is the subset of B given as $\{f(a) \mid a \in A\}$, i.e., the set of elements of B that arise as outputs of the function.

Functions of one Variable

In the context of functions of one variable, the term *function* is used for a function whose domain is a subset of $\mathbb{R}$ and whose co-domain is $\mathbb{R}$, i.e., for a real-valued function with a real variable as input. In other words, the term *function* is used for a function if both the domain and the range are subsets of $\mathbb{R}$.

Key Features

Equal Inputs give Equal Outputs

A key feature of functions is that, for a given function, the input to the function completely determines the output, i.e., if the same input is fed into the same function at different times, the output will be the same each time. This feature can be captured by the phrase equal inputs give equal outputs. When people ask whether a function is well defined, what they usually mean is whether it is a function at all, in the sense of whether it has the key feature of equal inputs give equal outputs.

Here are some examples:

Example of a purported function	Purported domain	Does it satisfy the *equal inputs give equal outputs* property?	So, is it *really* a function?
$f(d)$ is defined as the area of a circle with diameter d	Positive reals	Yes, because any two circles with the same diameter are congruent and hence have the same area. In fact, we can also obtain an *explicit* formula: $f(d) := \pi d^2 / 4$	Yes
$f(x)$ is defined as the area of a square with perimeter x	Positive reals	Yes, because any two squares with the same perimeter are congruent and hence have the same area. In fact, we can also obtain an *explicit* formula: $f(x) := x^2 / 16$	Yes
$f(x)$ is defined as the area of a rectangle with perimeter x	Positive reals	No, because different rectangles with the same perimeter can have different shapes and consequently have different areas	No

Domain and Restriction of Domain

The study of a function depends crucially on the domain on which the function is being studied. If a function of one variable is defined solely by means of an expression or procedure, the domain of the function is taken to be the largest possible subset of the reals on which that expression or procedure makes sense and gives a valid answer. However, we can also consider functions restricted to domains that are strictly smaller than the maximum possible domain on which the expression being used for the function makes sense. The behavior of the function, as well as answers to questions like whether it is increasing or decreasing and what its extreme values are, depends on what domain we are considering the function on.

Here are some examples: Consider the function f defined as follows: $f(d)$ *is defined as the area of a circle with diameter* d. Ignoring the boundary case of point circles and line circles, the only possible inputs for this function are positive reals, so f is a function from the positive reals to the positive reals given by the expression $f(d) := \pi d^2 / 4$. However, if we look *only* at the expression for f, then that expression makes sense for *all* real numbers, including zero and negative real numbers as well as positive real numbers. Call the latter function g, i.e., $g(d) := \pi d^2 / 4$ for all $d \in \mathbb{R}$. Then, f is the restriction of g to the subdomain $(0, \infty)$. Note that:

- g is not an increasing function, but the restriction f is an increasing function.
- g is not a one-one function, but the restriction f is a one-one function.
- g attains its absolute minimum value, but the restriction f does not.

Description of Functions

A *description* of a function should be a clear and actionable way of describing (i) what the domain is, and (ii) how to compute the output of the function from any given input in the domain. There are three main kinds of descriptions:

- Algebraic (expression-based or formula-based, or procedural).
- Numerical (in the form of a table of inputs and outputs).

- Graphical (in the form of the graph of the function).

Algebraic, Expression-based or Procedural Descriptions

This is a description of the function that uses a formula or expression based on known functions and known techniques for pointwise combination, composition, taking the inverse function, and using piecewise definitions. The known techniques are the tools used to build complicated functions from the existing simpler ones.

An algebraic description is typically written in the form:

$x \mapsto$ expression in x

The way this is interpreted is that, to evaluate the function at any actual input, we replace all occurrences of x on the right side with that input and compute.

Alternatively, if the function is denoted by the letter g, we can also write, in place of the above:

$g(x) :=$ expression in x

Dummy Variable: The letter x used above is a dummy variable (or a *local* variable). In other words, replacing x on both the left side and in the right side expression by a single other letter gives an *identical* function definition. Also, if the letter x is *already* in use, then some other letter should be used for the function description.

In some cases, the algebraic description may be too complicated to write in a single straight line expression. In this case, we may express it in terms of a procedure or algorithm.

Here are some examples:

Function	Algebraic description of function	Comments
identity function	$x \mapsto x$	
square function	$x \mapsto x^2$	
sine-squared function	$x \mapsto \sin^2 x$	We are using a composite of two functions: the square function and the sine function.
absolute value function	Failed to parse (syntax error): x \mapsto \left\lbrace x, & x \ge 0 \\ -x, & x < 0 \\\end{array}\right.	we use a piecewise definition of function. Specifically, this function is a piecewise linear function.

LIMIT OF A FUNCTION

The limit of a function at a point a in its domain (if it exists) is the value that the function approaches as its argument approaches a. The concept of a limit is the fundamental concept of calculus and analysis. It is used to define the derivative and the definite integral, and it can also be used to analyze the local behavior of functions near points of interest.

Informally, a function is said to have a limit L at a if it is possible to make the function arbitrarily close to L by choosing values closer and closer to a. Note that the actual value at a is irrelevant to the value of the limit.

The notation is as follows:

$$\lim_{x \to a} f(x) = L,$$

which is read as "the limit of $f(x)$ as xx approaches aa is L."

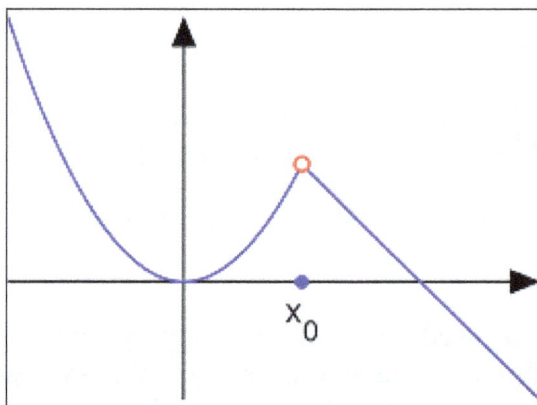

The limit of $f(x)$ at x_0 is the y-coordinate of the red point, not $f(x_0)$.

Formal Definition of a Function Limit

The limit of $f(x)$ as xx approaches x_0 is L, i.e

$$\lim_{x \to x_0} f(x) = L$$

if, for every $\in > 0$, there exists $\delta > 0$ such that, for all x,

$$0 < |x - x_0| < \delta \Rightarrow |f(x) - L| < \epsilon.$$

In practice, this definition is only used in relatively unusual situations. For many applications, it is easier to use the definition to prove some basic properties of limits and to use those properties to answer straightforward questions involving limits.

Properties of Limits

The most important properties of limits are the *algebraic* properties, which say essentially that limits respect algebraic operations.

Theorem

Suppose that $\lim_{x \to a} f(x) = M$ and $\lim_{x \to a} g(x) = N$.

Then
$$\lim_{x \to a} \left(f(x) + g(x) \right) = M + N$$

$$\lim_{x \to a} \left(f(x) - g(x) \right) = M - N$$

$$\lim_{x \to a} \left(f(x) \, g(x) \right) = M \, N$$

$$\lim_{x \to a} \left(\frac{f(x)}{g(x)} \right) = \frac{M}{N} \quad (\text{if } N \neq 0)$$

$$\lim_{x \to a} f(x)^k = M^k \quad (\text{if } M, k > 0)$$

These can all be proved via application of the epsilon-delta definition. Note that the results are only true if the limits of the individual functions exist: if $\lim_{x \to a} f(x)$ and $\lim_{x \to a} g(x)$ do not exist, the limit of their sum (or difference, product, or quotient) might nevertheless exist.

Coupled with the basic limits $\lim_{x \to a} c = c$, where cc is a constant, and $\lim_{x \to a} x = a$, the properties can be used to deduce limits involving rational functions:

Let $f(x)$ and $g(x)$ be polynomials, and suppose $g(a) \neq 0$. Then

$$\lim_{x \to a} \frac{f(x)}{g(x)} = \frac{f(a)}{g(a)}.$$

This is an example of continuity, or what is sometimes called limits by substitution.

$g(a) = 0$ is a more difficult case;

Example:

Let m and n be positive integers. Find

$$\lim_{x \to 1} \frac{x^m - 1}{x^n - 1}$$

Immediately substituting $x=1$ does not work, since the denominator evaluates to 0.0. First, divide top and bottom by $x-1$ to get

$$\frac{x^{m-1} + x^{m-2} + \cdots + 1}{x^n - 1 + x^{n-2} + \cdots + 1}.$$

Plugging in $x=1$ to the denominator does not give 0,0, so the limit is this fraction evaluated at $x=1$, which is

$$\frac{x^{m-1} + x^{m-2} + \cdots + 1}{x^{n-1} + x^{n-2} + \cdots + 1} = \frac{m}{n}.$$

It is important to notice that the manipulations in the above example are justified by the fact that $\lim_{x \to a} f(x)$ is independent of the value of $f(x)$ at $x = a$, or whether that value exists. This justifies for instance, dividing the top and bottom of the fraction $\dfrac{x^m - 1}{x^n - 1}$ by, $x-1$, since this is nonzero for $x \neq 1$.

One-sided Limits

A one-sided limit only considers values of a function that approaches a value from either above or below.

The right-side limit of a function f as it approaches a is the limit:

$$\lim_{x \to a} f(x) = L$$

The left-side limit of a function f is:

$$\lim_{x \to a^-} f(x) = L$$

The notation" $x \to a^{-n}$ indicates that we only consider values of x that are less than aa when evaluating the limit. Likewise, for " $x \to a^+$ "we consider only values greater than a. One-sided limits are important when evaluating limits containing absolute values $|x|$ sign (x), floor functions $\lfloor x \rfloor$ and other piecewise functions.

Example:

Find the left- and right-side limits of the signum function (x) as $x \to 0$:

$$\text{sgn}(x) = \begin{cases} \dfrac{|x|}{x} & x \neq 0 \\ 0 & x = 0. \end{cases}$$

Consider the following graph:

From this we see $\lim\limits_{x \to 0^+} \text{sgn}(x) = -1$.

Example:

Determine the limit $\lim\limits_{x \to 1^-} \dfrac{\sqrt{2x}(x-1)}{|x-1|}$

For $x < 1, |x-1|$ can be written as $-(x-1)$ Hence, the limit is $\lim\limits_{x \to 1^-} \dfrac{\sqrt{2x}(x-1)}{-(x-1)} = -\sqrt{2}$.

Two-sided Limits

By definition, a two-sided limit

$$\lim_{x \to a} f(x) = L$$

exists if the one-sided limits $\lim\limits_{x \to a^+} f(x)$ and $\lim\limits_{x \to a} f(x)$ are the same.

Example:

Compute the limit,

$$\lim_{x \to 1} \frac{|x-1|}{x-1}.$$

Since the absolute value function $f(x) = |x|$ is defined in a piecewise manner, we have to consider two limits: $\lim_{x \to 1} \dfrac{|x-1|}{x-1}$ and $\lim_{x \to 1^-} \dfrac{|x-1|}{x-1}$.

Start with the limit $\lim_{x \to 1^+} \dfrac{|x-1|}{x-1}$. For $x > 1, |x-1| = x-1$. So

$$\lim_{x \to 1^+} \frac{|x-1|}{x-1} = \lim_{x \to 1^+} \frac{x-1}{x-1} = 1.$$

Let us now consider the left-hand limit,

$$\lim_{x \to 1^-} \frac{|x-1|}{x-1}.$$

For $x < 1, x-1 = -|x-1|$. So

$$\lim_{x \to 1^-} \frac{|x-1|}{-|x-1|} = -1$$

So the two-sided limit $\lim_{x \to 1} \dfrac{|x-1|}{x-1}$ does not exist.

Example:

A graph of a function $f(x)$. As shown, it is continuous for all points except $x=-1$ and $x=2$ which are its asymptotes. Find all the integer points $-4 < I < 4$, where the two-sided limit $\lim_{x \to I} f(x)$ exists.

Since the graph is continuous at all points except $x=-1$ and $x=2$, the two-sided limit exists at $x = -3, x = -2, x = 0, x = 1$, and $x = 3$. At $x = 2$, there is no finite value for either of the two-sided limits, since the function increases without bound as the xx-coordinate approaches. The situation is similar for $x=-1$. So the points $x = -3, x = -2, x = 0, x = 1$, and $x = 3$ are all the integers on which two-sided limits are defined.

Infinite Limits

One way for a limit not to exist is for the one-sided limits to disagree. Another common way for a limit to not exist at a point aa is for the function to "blow up" near a, i.e. the function increases without bound. This happens in the above example at, $x=2$, where there is a vertical asymptote. This common situation gives rise to the following notation.

Given a function $f(x)$ and a real number a, a, we say

$$f(x) = \infty$$

If the function can be made arbitrarily large by moving x sufficiently close to a,

for all $N > 0$, there exists $\delta > 0$ such that $0 < |x - a| < \delta \Rightarrow f(x) > N$.

There are similar definitions for one-sided limits, as well as limits "approaching $-\infty$."

Warning: If $\lim\limits_{x \to a} f(x) = \infty$, it is tempting to say that the limit at a exists and equals ∞. This is incorrect. If $\lim\limits_{x \to a} f(x) = \infty$, the limit does not exist; the notation merely gives information about the way in which the limit fails to exist, i.e. the value of the function "approaches ∞" or increases without bound as $x \to a$.

Example:

What can we say about $\lim\limits_{x \to 0} \dfrac{1}{x}$?

Separating the limit into $\lim\limits_{x \to 0^+} \frac{1}{x}$ and $\lim\limits_{x \to 0^-} \frac{1}{x}$ we obtain $\lim\limits_{x \to 0^+} \dfrac{1}{x} = \infty$

and

$$\lim\limits_{x \to 0^-} \dfrac{1}{x} = -\infty$$

To prove the first statement, for any $N > 0$ in the formal definition, we can take $\delta = \dfrac{1}{N}$, and the proof of the second statement is similar.

So the function increases without bound on the right side and decreases without bound on the left side. We cannot say anything else about the two-sided limit $\lim\limits_{x \to a} \dfrac{1}{x} \neq \infty$ or.

Example:

What can we say about $\lim\limits_{x \to 0} \dfrac{1}{x^2}$?

Separating the limit into $\lim\limits_{x \to 0^+} \dfrac{1}{x}$ and $\lim\limits_{x \to 0^-} \dfrac{1}{x}$, we obtain

$$\lim\limits_{x \to 0^+} \dfrac{1}{x^2} = \infty$$

and

$$\lim\limits_{x \to 0^-} \dfrac{1}{x^2} = \infty.$$

Since these limits are the same, we have $\lim_{x\to 0}\dfrac{1}{x^2}=\infty$. Again, this limit does not, strictly speaking, exist, but the statement is meaningful nevertheless, as it gives information about the behavior of the function $\dfrac{1}{x^2}$ near 0.

Limits at Infinity

Another extension of the limit concept comes from considering the function's behavior as x "approaches ∞," that is, as x increases without bound.

The equation $\lim_{x\to\infty} f(x)=L$ means that the values of f can be made arbitrarily close to L by taking x sufficiently large. That is,

for all $\epsilon > 0$, there is $N > 0$ such that $x > N \implies |f(x)-L| < \epsilon$.

There are similar definitions for $\lim_{x\to -\infty} f(x)=L$, as well as $\lim_{x\to -\infty} f(x)=\infty$ and so on.

Graphically, $\lim_{x\to a} f(x)=\infty$ corresponds to a vertical asymptote at a, while $\lim_{x\to\infty} f(x)=L$ corresponds to a horizontal asymptote at L.

Limits by Factoring

Limits by factoring refers to a technique for evaluating limits that requires finding and eliminating common factors.

Limits by Substitution

Evaluating limits by substitution refers to the idea that under certain circumstances (namely if the function we are examining is continuous), we can evaluate the limit by simply evaluating the function at the point we are interested in.

L'Hôpital's Rule

L'Hôpital's rule is an approach to evaluating limits of certain quotients by means of derivatives.

Specifically, under certain circumstances, it allows us to replace $\lim \dfrac{f(x)}{g(x)}$ with $\lim \dfrac{f'(x)}{g'(x)}$ which is frequently easier to evaluate.

CONTINUITY OF A FUNCTION

The property of continuity is exhibited by various aspects of nature. The water flow in the rivers is continuous. The flow of time in human life is continuous i.e. you are getting older continuously. And so on. Similarly, in mathematics, we have the notion of the continuity of a function.

What it simply means is that a function is said to be continuous if you can sketch its curve on a graph without lifting your pen even once. It is a very straightforward and close to accurate definition actually. But for the sake of higher mathematics, we must define it in a more precise way.

A function f(x) is said to be continuous at a point x = a, in its domain if the following three conditions are satisfied:

1.   f(a) exists (i.e. the value of f(a) is finite).

2. $\text{Lim}_{x \to a}$ f(x) exists (i.e. the right-hand limit = left-hand limit, and both are finite).

3. $\text{Lim}_{x \to a}$ f(x) = f(a).

The function f(x) is said to be continuous in the interval $I = [x_1, x_2]$ if the three conditions mentioned above are satisfied for every point in the interval I.

However, note that at the end-points of the interval I, we need not consider both the right-hand and the left-hand limits for the calculation of $\text{Lim}_{x \to a} f(x)$. For $a = x_1$, only the right-hand limit need be considered, and for a = x_2, only the left-hand limit needs to be considered.

Some Typical Continuous Functions

- Trigonometric Functions in certain periodic intervals (sin x, cos x, tan x etc.)

- Polynomial Functions ($x^2 + x + 1$, $x^4 + 2$ ....etc.)

- Exponential Functions (e^{2x}, $5e^x$ etc.)

- Logarithmic Functions in their domain ($\log_{10} x$, ln x^2 etc.)

Discontinuity

If any one of the three conditions for a function to be continuous fails; then the function is said to be discontinuous at that point. On the basis of the failure of which specific condition leads to discontinuity, we can define different types of discontinuities.

In this type of discontinuity, the right-hand limit and the left-hand limit for the function at x = a exists; but the two are not equal to each other. It can be shown as:

$$Lim_{x \to a} + f(x) \neq Lim_{x \to a-} f(x)$$

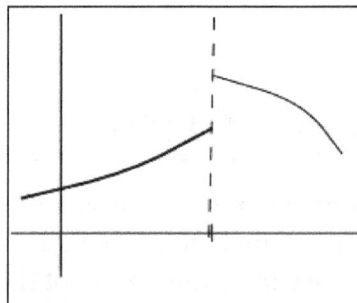

Jump

Infinite Discontinuity

The function diverges at x = a to give it a discontinuous nature here. That is to say, f(a) is not defined. Since the value of the function at x = a tends to infinity or doesn't approach a particular finite value, the limits of the function as x → a are also not defined.

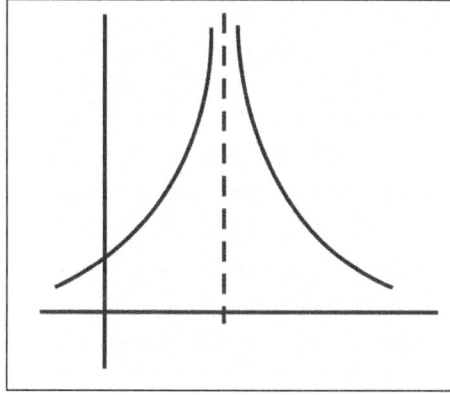

Infinite

Point Discontinuity

This is a category of discontinuity in which the function has a well defined two-sided limit at x = a, but either f(a) is not defined or f(a) is not equal to its limit. The discrepancy can be shown as:

$$Lim_{x \to a} f(x) \neq f(a)$$

This type of discontinuity is also known as a Removable Discontinuity since it can be easily eliminated by redefining the function in such a way that,

$$f(a) = Lim_{x \to a} f(x)$$

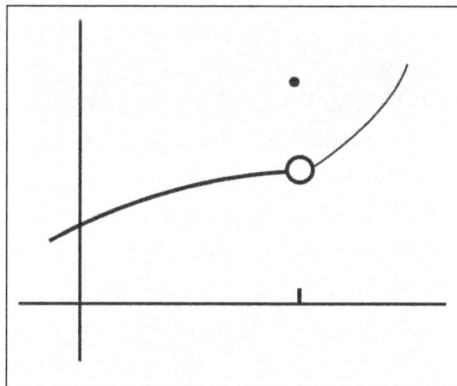

Removable

Example: Let a function be defined as

$$f(x) = 5 - 2x \text{ for } x < 1$$
$$3 \text{ for } x = 1$$
$$x + 2 \text{ for } x > 1$$

Is this function continuous for all x?

Solution: Since for x < 1 and x > 1, the function f(x) is defined by straight lines (that can be drawn continuously on a graph), the function will be continuous for all $x \neq 1$. Now for x = 1, let us check all the three conditions:

$$-> f(1) = 3 \ (\text{given})$$

$$-> \ \text{Left-Hand Limit}$$
$$= Lim_{x \to 1^-} f(x)$$
$$= Lim_{x \to 1^-} (5 - 2x)$$
$$= 5 - 2 \times 1$$
$$= 3$$

$$-> \ \text{Right-Hand Limit}$$
$$= Lim_{x \to 1^+} f(x)$$
$$= Lim_{x \to 1^+} (x + 2)$$
$$= 1 + 2$$
$$= 3$$

$$-> Lim_{x \to 1^-} f(x) = Lim_{x \to 1} + f(x) = 3 = f(1)$$

Thus all the three conditions are satisfied and the function f(x) is found out to be continuous at x = 1. Therefore, f(x) is continuous for all x.

Differential and Integral Calculus

Differential calculus is the branch of calculus which deals with the rate of change in quantities while integral calculus focuses on the accumulation of quantities and the areas under and between curves. Some of the types of integrals studied under this discipline are quadratic integrals, Borwein integrals and Dirichlet integrals. The diverse applications of differential and integral calculus have been thoroughly discussed in this chapter.

DIFFERENTIAL CALCULUS

Differential calculus is a branch of mathematics dealing with the concepts of derivative and differential and the manner of using them in the study of functions. The development of differential calculus is closely connected with that of integral calculus. Indissoluble is also their content. Together they form the base of mathematical analysis, which is extremely important in the natural sciences and in technology. The introduction of variable magnitudes into mathematics by R. Descartes was the principal factor in the creation of differential calculus. Differential and integral calculus were created, in general terms, by I. Newton and G. Leibniz towards the end of the 17th century, but their justification by the concept of limit was only developed in the work of A.L. Cauchy in the early 19th century. The creation of differential and integral calculus initiated a period of rapid development in mathematics and in related applied disciplines. Differential calculus is usually understood to mean classical differential calculus, which deals with real-valued functions of one or more real variables, but its modern definition may also include differential calculus in abstract spaces. Differential calculus is based on the concepts of real number; function; limit and continuity — highly important mathematical concepts, which were formulated and assigned their modern content during the development of mathematical analysis and during studies of its foundations. The central concepts of differential calculus — the derivative and the differential — and the apparatus developed in this connection furnish tools for the study of functions which locally look like linear functions or polynomials, and it is in fact such functions which are of interest, more than other functions, in applications.

Let a function $y = f(x)$ be defined in some neighbourhood of a point x_0. Let $\Delta x \neq 0$ denote the increment of the argument and let $\Delta y = f(x_0 + \Delta x) - f(x_0)$ denote the corresponding increment of the value of the function. If there exists a (finite or infinite) limit:

$$\lim_{\Delta x \to 0} \frac{\Delta y}{\Delta x}$$

then this limit is said to be the derivative of the function f at x_0; it is denoted by $f'(x_0)$, $df(x_0)/dx$, y', y'_x, dy/dx. Thus, by definition,

$$f'(x_0) = \lim_{\Delta x \to 0} \frac{\Delta y}{\Delta x} = \lim_{\Delta x \to 0} \frac{f(x_0 + \Delta x) - f(x_0)}{\Delta x}$$

The operation of calculating the derivative is called differentiation. If $f'(x_0)$ is finite, the function f is called differentiable at the point x_0. A function which is differentiable at each point of some interval is called differentiable in the interval.

Geometric Interpretation of the Derivative

Let C be the plane curve defined in an orthogonal coordinate system by the equation $y = f(x)$ where f is defined and is continuous in some interval J; let $M(x_0, y_0)$ be a fixed point on C, let $P(x, y)$ ($x \in J$) be an arbitrary point of the curve C and let MP be the secant. An oriented straight line MT (T a variable point with abscissa $x_0 + \Delta x$) is called the tangent to the curve C at the point M if the angle ϕ between the secant MP and the oriented straight line tends to zero as $x \to x_0$ (in other words, as the point $P \in C$ arbitrarily tends to the point M). If such a tangent exists, it is unique. Putting $x = x_0 + \Delta x$, $\Delta y = f(x_0 + \Delta x) - f(x_0)$, one obtains the equation $\tan \beta = \Delta y / x$ for the angle β between MP and the positive direction of the x-axis.

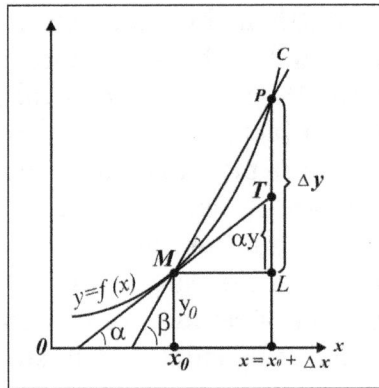

The curve C has a tangent at the point M if and only if $\lim_{Dx \to 0} \Delta y / \Delta x$ exists, i.e. if $f'(x_0)$ exists. The equation $\tan \alpha = f'(x_0)$ is valid for the angle α between the tangent and the positive direction of the x-axis. If $f'(x_0)$ is finite, the tangent forms an acute angle with the positive x-axis, i.e. $-\pi/2 < \alpha < \pi/2$; if $f'(x_0) = \infty$, the tangent forms a right angle with that axis.

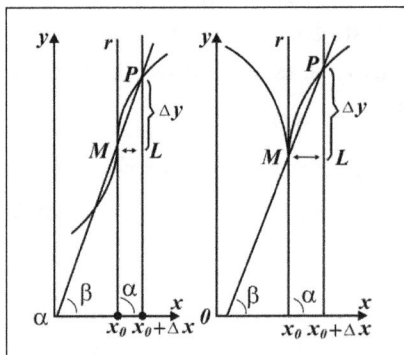

Thus, the derivative of a continuous function f at a point x_0 is identical to the slope $\tan \alpha$ of the tangent to the curve defined by the equation $y = f(x)$ at its point with abscissa x_0.

Mechanical Interpretation of the Derivative

Let a point M move in a straight line in accordance with the law $s = f(t)$. During time Δt the point M becomes displaced by $\Delta s = f(t + \Delta t) - f(t)$. The ratio $\Delta s / \Delta t$ represents the average velocity v_{av} during the time Δt. If the motion is non-uniform, v_{av} is not constant. The instantaneous velocity at the moment t is the limit of the average velocity as $\Delta t \to 0$, i.e. $v = f'(t)$ (on the assumption that this derivative in fact exists).

Thus, the concept of derivative constitutes the general solution of the problem of constructing tangents to plane curves, and of the problem of calculating the velocity of a rectilinear motion. These two problems served as the main motivation for formulating the concept of derivative.

A function which has a finite derivative at a point x_0 is continuous at this point. A continuous function need not have a finite nor an infinite derivative. There exist continuous functions having no derivative at any point of their domain of definition.

The formulas given below are valid for the derivatives of the fundamental elementary functions at any point of their domain of definition (exceptions are stated):

1. if $f(x) = C = \text{const}$, then $f'(x) = C' = 0$;

2. if $f(x) = x$, then $f'(x) = 1$;

3. $(x^\alpha)' = \alpha x^{\alpha-1}$, $\alpha = \text{const}$ ($x \neq 0$, if $\alpha \leq 1$);

4. $(\alpha^x)' = \alpha^x \ln a$, $a = \text{const} > 0$, $a \neq 1$; in particular, $(e^x)' = e^x$;

5. $(\log_a x)' = (\log_a e) / x = 1 / (x \ln \alpha)$, $a = \text{const} > 0$, $a \neq 1$, $(\ln x)' = 1 / x$;

6. $(\sin x)' = \cos x$;

7. $(\cos x)' = -\sin x$;

8. $(\tan x)' = 1 / \cos^2 x$;

9. $(\cotan x)' = -1 / \sin^2 x$;

10. $(\arcsin x)' = 1 / \sqrt{1 - x^2}$, $x \neq \pm 1$;

11. $(\arccos x)' = -1 / \sqrt{1 - x^2}$, $x \neq \pm 1$;

12. $(\arctan x)' = 1 / (1 + x^2)$;

13. $(\arccotan x)' = -1 / (1 + x^2)$;

14. $(\sinh x)' = \cosh x$;

15. $(\cosh x)' = \sinh x$;

16. $(\tanh x)' = 1/\cosh^2 x;$

17. $(\coth x)' = -1/\sinh^2 x.$

The following laws of differentiation are valid:

If two functions u and v are differentiable at a point x_0, then the functions

$$cu \quad (\text{where } c = \text{const}), \quad u \pm v, \quad uv, \quad \frac{u}{v}(v \neq 0)$$

are also differentiable at that point, and

$$(cu)' = cu',$$
$$(u \pm v)' = u' \pm v',$$
$$(uv)' = u'v + uv'$$
$$\left(\frac{u}{v}\right)' = \frac{u'v - uv'}{v^2}.$$

Theorem on the derivative of a composite function: If the function $y = f(u)$ is differentiable at a point u_0, while the function $\phi(x)$ is differentiable at a point x_0, and if $u_0 = \phi(x_0)$, then the composite function $y = f(\phi(x))$ is differentiable at x_0, and $y'_x = f'(u_0)\phi'(x_0)$ or, using another notation, $dy/dx = (dy/du)(du/dx)$.

Theorem on the derivative of the inverse function: If $y = f(x)$ and $x = g(y)$ are two mutually inverse increasing (or decreasing) functions, defined on certain intervals, and if $f'(x_0) \neq 0$ exists (i.e. is not infinite), then at the point $y_0 = f(x_0)$ the derivative $g'(y_0) = 1/f'(x_0)$ exists, or, in a different notation, $dx/dy = 1/(dy/dx)$. This theorem may be extended: If the other conditions hold and if also $f'(x_0) = 0$ or $f'(x_0) = \infty$, then, respectively, $g'(y_0) = \infty$ or $g'(y_0) = 0$.

One-sided Derivatives

If at a point x_0 the limit:

$$\lim_{\Delta x \downarrow 0} \frac{\Delta y}{\Delta x}$$

exists, it is called the right-hand derivative of the function $y = f(x)$ at x_0 (in such a case the function need not be defined everywhere in a certain neighbourhood of the point x_0; this requirement may then be restricted to $x \geq x_0$). The left-hand derivative is defined in the same way, as:

$$\lim_{\Delta x \uparrow 0} \frac{\Delta y}{\Delta x}.$$

A function f has a derivative at a point x_0 if and only if equal right-hand and left-hand derivatives exist at that point. If the function is continuous, the existence of a right-hand (left-hand) derivative at a point is equivalent to the existence, at the corresponding point of its graph, of a right (left) one-sided semi-tangent with slope equal to the value of this one-sided derivative. Points at which the semi-tangents do not form a straight line are called angular points or cusps.

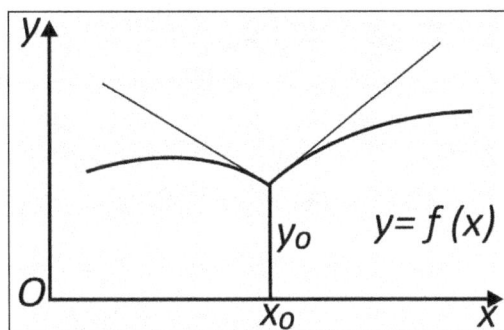

Derivatives of Higher Orders

Let a function $y = f(x)$ have a finite derivative $y' = f'(x)$ at all points of some interval; this derivative is also known as the first derivative, or the derivative of the first order, which, being a function of x, may in its turn have a derivative $y'' = f''(x)$, known as the second derivative, or the derivative of the second order, of the function f, etc. In general, the n-th derivative, or the derivative of order n, is defined by induction by the equation $y^{(n)} = (y^{(n-1)})'$, on the assumption that $y^{(n-1)}$ is defined on some interval. The notations employed along with $y^{(n)}$ are $f^{(n)}$, $d^n f(x)/dx^n$, and, if $n = 2,3$, also y'', $f''(x)$, y''', $f'''(x)$.

The second derivative has a mechanical interpretation: It is the acceleration $w = d^2 s/dt^2 = f''(t)$ of a point in rectilinear motion according to the law $s = f(t)$.

Differential

Let a function $y = f(x)$ be defined in some neighbourhood of a point x and let there exist a number A such that the increment Δy may be represented as $\Delta y = A\Delta x + \omega$ with $\omega/\Delta x \to 0$ as $\Delta x \to 0$. The term $A\Delta x$ in this sum is denoted by the symbol dy or df and is named the differential of the function $f(x)$ (with respect to the variable x) at x. The differential is the principal linear part of increment of the function (its geometrical expression is the segment LT in figure, where MT is the tangent to $y = f(x)$ at the point (x_0, y_0) under consideration).

The function $y = f(x)$ has a differential at x if and only if it has a finite derivative:

$$f'(x) = \lim_{\Delta x \to 0} \frac{\Delta y}{\Delta x} = A$$

At this point, a function for which a differential exists is called differentiable at the point in question. Thus, the differentiability of a function implies the existence of both the differential and the finite derivative, and $dy = df(x) = f'(x)\Delta x$. For the independent variable x one puts $dx = \Delta x$, and one may accordingly write $dy = f'(x)dx$, i.e. the derivative is equal to the ratio of the differentials:

$$f'(x) = \frac{dy}{dx}$$

The formulas and the rules for computing derivatives lead to corresponding formulas and rules for calculating differentials. In particular, the theorem on the differential of a composite function is valid: If a function $y = f(u)$ is differentiable at a point u_0, while a function $\phi(x)$ is differentiable at a point x_0 and $u_0 = \phi(x_0)$, then the composite function $y = f(\phi(x))$ is differentiable at the point x_0 and $dy = f'(u_0)\,du$, where $du = \phi'(x_0)\,dx$. The differential of a composite function has exactly the form it would have if the variable u were an independent variable. This property is known the invariance of the form of the differential. However, if u is an independent variable, $du = \Delta u$ is an arbitrary increment, but if u is a function, du is the differential of this function which, in general, is not identical with its increment.

Differentials of Higher Orders

The differential dy is also known as the first differential, or differential of the first order. Let $y = f(x)$ have a differential $dy = f'(x)dx$ at each point of some interval. Here $dx = \Delta x$ is some number independent of x and one may say, therefore, that $dx = \text{const}$. The differential dy is a function of x alone, and may in turn have a differential, known as the second differential, or the differential of the second order, of f, etc. In general, the n-th differential, or the differential of order n, is defined by induction by the equality $d^n y = d(d^{n-1}y)$, on the assumption that the differential $d^{n-1}y$ is defined on some interval and that the value of dx is identical at all steps. The invariance condition for $d^2y, d^3y, \ldots$, is generally not satisfied (with the exception $y = f(u)$ where u is a linear function).

The repeated differential of dy has the form,

$$\delta(dy) = f''(x)dx\delta x$$

and the value of $\delta(dy)$ for $dx = \delta x$ is the second differential.

Principal Theorems and Applications of Differential Calculus

The fundamental theorems of differential calculus for functions of a single variable are usually considered to include the Rolle theorem, the Legendre theorem (on finite variation), the Cauchy theorem, and the Taylor formula. These theorems underlie the most important applications of differential calculus to the study of properties of functions — such as increasing and decreasing functions, convex and concave graphs, finding the extrema, points of inflection, and the asymptotes of a graph. Differential calculus makes it possible to compute the limits of a function in many cases when this is not feasible by the simplest limit theorems. Differential calculus is extensively applied in many fields of mathematics, in particular in geometry.

Differential Calculus of Functions in Several Variables

Let a function $z = f(x, y)$ be given in a certain neighbourhood of a point (x_0, y_0) and let the value $y = y_0$ be fixed. $f(x, y_0)$ will then be a function of x alone. If it has a derivative with respect to x at x_0, this derivative is called the partial derivative of f with respect to x at (x_0, y_0); it is denoted by $f_x'(x_0, y_0)$, $\partial f(x_0, y_0)/\partial x$, $\partial f/\partial x$, z_x', $\partial z/\partial x$, or $f_x(x_0, y_0)$. Thus, by definition,

$$f_x'(x_0, y_0) = \lim_{\Delta x \to 0} \frac{\Delta_x z}{\Delta x} = \lim_{\Delta x \to 0} \frac{f(x_0 + \Delta x, y_0) - f(x_0, y_0)}{\Delta x}$$

where $\Delta_x z = f(x_0 + \Delta x, y_0) - f(x_0, y_0)$ is the partial increment of the function with respect to x (in the general case, $\partial z / \partial x$ must not be regarded as a fraction; $\partial / \partial x$ is the symbol of an operation).

The partial derivative with respect to y is defined in a similar manner:

$$f_y'(x_0, y_0) = \lim_{\Delta y \to 0} \frac{\Delta_y z}{\Delta y} = \lim_{\Delta y \to 0} \frac{f(x_0, y_0 + \Delta y) - f(x_0, y_0)}{\Delta y}$$

where $\Delta_y z$ is the partial increment of the function with respect to $\Delta_y z$. Other notations include $\partial f(x_0, y_0) / \partial y$, $\partial f / \partial y$, z_y', $\partial z / \partial y$, and $f_y'(x_0, y_0)$. Partial derivatives are calculated according to the rules of differentiation of functions of a single variable (in computing z_x' one assumes $y = \text{const}$ while if z_y' is calculated, one assumes $x = \text{const}$).

The partial differentials of $z = f(x, y)$ at (x_0, y_0) are, respectively,

$$d_x z = f_x'(x_0, y_0) dx; \qquad d_y z = f_y'(x_0, y_0) dy,$$

where, as in the case of a single variable, $dx = \Delta x, dy = \Delta y$ denote the increments of the independent variables.

The first partial derivatives $\partial z / \partial x = f_x'(x, y)$ and $\partial z / \partial y = f_y'(x, y)$, or the partial derivatives of the first order, are functions of x and y, and may in their turn have partial derivatives with respect to x and y. These are named, with respect to the function $z = f(x, y)$, the partial derivatives of the second order, or second partial derivatives. It is assumed that

$$\frac{\partial}{\partial x}\left(\frac{\partial z}{\partial x}\right) = \frac{\partial^2 z}{\partial x^2}, \quad \frac{\partial}{\partial y}\left(\frac{\partial z}{\partial x}\right) = \frac{\partial^2 z}{\partial x \partial y}$$

$$\frac{\partial}{\partial x}\left(\frac{\partial z}{\partial y}\right) = \frac{\partial^2 z}{\partial y \partial x}, \quad \frac{\partial}{\partial y}\left(\frac{\partial z}{\partial x}\right) = \frac{\partial^2 z}{\partial y^2}$$

The following notations are also used instead of $\partial^2 z / \partial^2 x$:

$$z_{xx}'' \quad z_{x^2}'' \quad \frac{\partial^2 f(x, y)}{\partial x^2}, \frac{\partial^2 f}{\partial x^2}, \quad f_{xx}''(x, y), \quad f_{x^2}''(x, y), \quad f_{xx}(x, y);$$

and instead of $\partial^2 z / \partial x \partial y$:

$$z_{xy}'', \quad \frac{\partial^2 f(x, y)}{\partial x \partial y}, \frac{\partial^2 f}{\partial x \partial y}, \quad f_{xy}''(x, y), \quad f_{xy}(x, y),$$

One can introduce in the same manner partial derivatives of the third and higher orders, together with the respective notations: $\partial^n z / \partial x^n$ means that the function z is to be differentiated n times with respect to x; $\partial^n z / \partial x^p \partial y^q$ where $n = p + q$ means that the function z is differentiated p times with respect to x and q times with respect to y. The partial derivatives of second and higher orders obtained by differentiation with respect to different variables are known as mixed partial derivatives.

To each partial derivative corresponds some partial differential, obtained by its multiplication by the differentials of the independent variables taken to the powers equal to the number of differentiations with respect to the respective variable. In this way one obtains the n-th partial differentials, or the partial differentials of order n:

$$n \quad \frac{\partial^n z}{\partial x^n} \, dx^n, \quad \frac{\partial^n z}{\partial x^p \partial y^q} \, dx^p \, dy^q.$$

The following important theorem on derivatives is valid: If, in a certain neighbourhood of a point (x_0, y_0), a function $z = f(x, y)$ has mixed partial derivatives $f''_{xy}(x, y)$ and $f''_{yx}(x, y)$, and if these derivatives are continuous at the point (x_0, y_0), then they coincide at this point.

A function $z = f(x, y)$ is called differentiable at a point (x_0, y_0) with respect to both variables x and y if it is defined in some neighbourhood of this point, and if its total increment

$$\Delta z = f(x_0 + \Delta x, y_0 + \Delta y) - f(x_0, y_0)$$

may be represented in the form

$$\Delta z = A\Delta x + B\Delta y + \omega,$$

where A and B are certain numbers and $\omega / \rho \to 0$ for $\rho = \sqrt{(\Delta x)^2 + (\Delta y)^2} \to 0$ (provided that the point $(x_0 + \Delta x, y_0 + \Delta y)$ lies in this neighbourhood). In this context, the expression

$$dz = df(x_0, y_0) = A \, \Delta x + B \, \Delta y$$

is called the total differential (of the first order) of f at (x_0, y_0); this is the principal linear part of increment. A function which is differentiable at a point is continuous at that point (the converse proposition is not always true). Moreover, differentiability entails the existence of finite partial derivatives:

$$f'_x(x_0, y_0) = \lim_{\Delta x \to 0} \frac{\Delta_x z}{\Delta x} = A, \quad f'_y(x_0, y_0) = \lim_{\Delta y \to 0} \frac{\Delta_y z}{\Delta y} = B.$$

Thus, for a function which is differentiable at (x_0, y_0),

$$dz = df(x_0, y_0) = f'_x(x_0, y_0) \, \Delta x + f'_y(x_0, y_0)\Delta y,$$

or

$$dz = df(x_0, y_0) = f'_x(x_0, y_0) \, dx + f'_y(x_0, y_0)dy,$$

if, as in the case of a single variable, one puts, for the independent variables, $dx = \Delta x$, $dy = \Delta y$.

The existence of finite partial derivatives does not, in the general case, entail differentiability (unlike in the case of functions in a single variable). The following is a sufficient criterion of the differentiability of a function in two variables: If, in a certain neighbourhood of a point

(x_0, y_0), a function f has finite partial derivatives f'_x and f'_y which are continuous at (x_0, y_0), then f is differentiable at this point. Geometrically, the total differential $df(x_0, y_0)$ is the increment of the applicate of the tangent plane to the surface $z = f(x, y)$ at the point (x_0, y_0, z_0), where $z_0 = f(x_0, y_0)$.

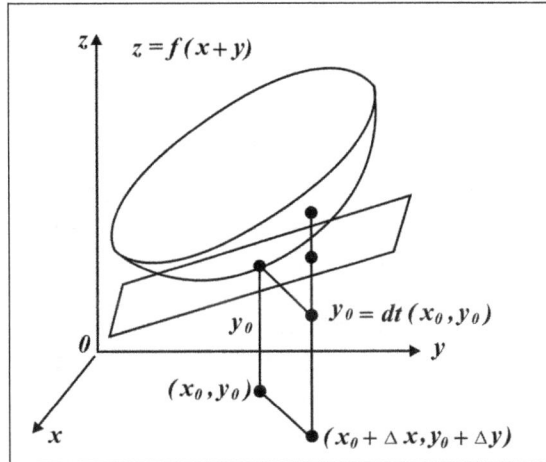

Total differentials of higher orders are, as in the case of functions of one variable, introduced by induction, by the equation

$$d^n z = d(d^{n-1} z),$$

on the assumption that the differential $d^{n-1} z$ is defined in some neighbourhood of the point under consideration, and that equal increments of the arguments dx, dy are taken at all steps. Repeated differentials are defined in a similar manner.

Derivatives and Differentials of Composite Functions

Let $w = f(u_1, \ldots, u_m)$ be a function in m variables which is differentiable at each point of an open domain D of the m-dimensional Euclidean space R^m, and let m functions $u_1 = \phi(x_1, \ldots, x_n), \ldots, u_m = \phi_m(x_1, \ldots, x_n)$ in n variables be defined in an open domain G of the n-dimensional Euclidean space R^n. Finally, let the point $(u_1, \ldots, u_m)$, corresponding to a point $(x_1, \ldots, x_n) \in G$, be contained in D. The following theorems then hold:

A) If the functions $\phi_1, \ldots, \phi_m$ have finite partial derivatives with respect to $x_1, \ldots, x_n$, the composite function $w = f(u_1, \ldots, u_m)$ in $x_1, \ldots, x_n$ also has finite partial derivatives with respect to $x_1, \ldots, x_n$, and

$$\frac{\partial w}{\partial x_1} = \frac{\partial f}{\partial u_1} \frac{\partial u_1}{\partial x_1} + \ldots + \frac{\partial f}{\partial u_n} \frac{\partial u_n}{\partial x_1},$$

$$\ldots \ldots \ldots \ldots \ldots$$

$$\frac{\partial w}{\partial x_n} = \frac{\partial f}{\partial u_n} \frac{\partial u_1}{\partial x_n} + \ldots + \frac{\partial f}{\partial u_n} \frac{\partial u_n}{\partial x_n},$$

B) If the functions $\phi_1, \ldots, \phi_m$ are differentiable with respect to all variables at a point $(x_1, \ldots, x_n) \in G$, then the composite function $w = f(u_1, \ldots, u_m)$ is also differentiable at that point, and

$$dw = \frac{\partial f}{\partial u_1} du_1 + \ldots + \frac{\partial f}{\partial u_n} du_n,$$

where $du_1, \ldots, du_m$ are the differentials of the functions $u_1, \ldots, u_m$. Thus, the property of invariance of the first differential also applies to functions in several variables. It does not usually apply to differentials of the second or higher orders.

Differential calculus is also employed in the study of the properties of functions in several variables: finding extrema, the study of functions defined by one or more implicit equations, the theory of surfaces, etc. One of the principal tools for such purposes is the Taylor formula.

The concepts of derivative and differential and their simplest properties, connected with arithmetical operations over functions and superposition of functions, including the property of invariance of the first differential, are extended, practically unchanged, to complex-valued functions in one or more variables, to real-valued and complex-valued vector functions in one or several real variables, and to complex-valued functions and vector functions in one or several complex variables. In functional analysis the ideas of the derivative and the differential are extended to functions of the points in an abstract space.

Derivative Test

In calculus, a derivative test uses the derivatives of a function to locate the critical points of a function and determine whether each point is a local maximum, a local minimum, or a saddle point. Derivative tests can also give information about the concavity of a function.

The usefulness of derivatives to find extrema is proved mathematically by Fermat's theorem of stationary points.

First Derivative Test

The first derivative test examines a function's monotonic properties (where the function is increasing or decreasing) focusing on a particular point in its domain. If the function "switches" from increasing to decreasing at the point, then the function will achieve a highest value at that point. Similarly, if the function "switches" from decreasing to increasing at the point, then it will achieve a least value at that point. If the function fails to "switch", and remains increasing or remains decreasing, then no highest or least value is achieved.

One can examine a function's monotonicity without calculus. However, calculus is usually helpful because there are sufficient conditions that guarantee the monotonicity properties above, and these conditions apply to the vast majority of functions one would encounter.

Precise Statement of Monotonicity Properties

Stated precisely, suppose f is a continuous real-valued function of a real variable, defined on some interval containing the point x.

- If there exists a positive number $r>0$ such that f is weakly increasing on $(x - r, x)$ and weakly decreasing on $(x, x + r)$, then f has a local maximum at x. This statement also works the other way around, if x is a local maximum point, then f is weakly increasing on $(x - r, x)$ and weakly decreasing on $(x, x + r)$.

- If there exists a positive number $r>0$ such that f is strictly increasing on $(x - r, x)$ and strictly increasing on $(x, x + r)$, then f is strictly increasing on $(x - r, x + r)$ and does not have a local maximum or minimum at x.

This statement is a direct consequence of how local extrema are defined. That is, if x_o is a local maximum point then there exists $r>0$ such that $f(x) \leq f(x_0)$ for x in $(x - r, x + r)$ which means f has to increase from $x - r$ to x and has to decrease from x to $x + r$ because f is continuous.

Note that in the first two cases, f is not required to be strictly increasing or strictly decreasing to the left or right of x, while in the last two cases, f is required to be strictly increasing or strictly decreasing. The reason is that in the definition of local maximum and minimum, the inequality is not required to be strict: e.g. every value of a constant function is considered both a local maximum and a local minimum.

Precise Statement of First Derivative Test

The first derivative test depends on the "increasing-decreasing test", which is itself ultimately a consequence of the mean value theorem. It is a direct consequence of the way the derivative is defined and its connection to decrease and increase of a function locally.

Suppose f is a real-valued function of a real variable defined on some interval containing the critical point a. Further suppose that f is continuous at a and differentiable on some open interval containing a, except possibly at a itself.

- If there exists a positive number $r>0$ such that for every x in $(a - r, a)$ we have $f'(x) \geq 0$, and for every x in $(a, a + r)$ we have $f'(x) \leq 0$, then f has a local maximum at a.

- If there exists a positive number $r>0$ such that for every x in $(a - r, a) \cup (a, a + r)$ we have $f'(x) > 0$, then f is strictly increasing at a and has neither a local maximum nor a local minimum there.

- If none of the above conditions hold, then the test fails. (Such a condition is not vacuous; there are functions that satisfy none of the first three conditions such as $f(x) = x^2 \cdot \sin(1/x)$.)

Again, note that in the first two cases, the inequality is not required to be strict, while in the next two, strict inequality is required.

Applications

The first derivative test is helpful in solving optimization problems in physics, economics, and engineering. In conjunction with the extreme value theorem, it can be used to find the absolute maximum and minimum of a real-valued function defined on a closed, bounded interval. In conjunction with other information such as concavity, inflection points, and asymptotes, it can be used to sketch the graph of a function.

Second Derivative Test (Single Variable)

After establishing the critical points of a function, the *second derivative test* uses the value of the second derivative at those points to determine whether such points are a local maximum or a local minimum. If the function f is twice differentiable at a critical point x (i.e. $f'(x) = o$), then:

- If $f''(x) < 0$ then f has a local maximum at x.

- If $f''(x) > 0$ then f has a local minimum at x.

- If $f''(x) = 0$, the test is inconclusive.

In the last case, Taylor's Theorem may be used to determine the behavior of f near x using higher derivatives.

Proof of the Second Derivative Test

Suppose we have $f''(x) > 0$ (the proof for $f''(x) < 0$ is analogous). By assumption, $f'(x) = 0$. Then,

$$0 < f''(x) = \lim_{h \to 0} \frac{f'(x+h) - f'(x)}{h} = \lim_{h \to 0} \frac{f'(x+h) - 0}{h} = \lim_{h \to 0} \frac{f'(x+h)}{h}$$

Thus, for h sufficiently small we get,

$$\frac{f'(x+h)}{h} > 0$$

which means that $f'(x+h) < 0$ if $h < o$ (intuitively, f is decreasing as it approaches x from the left), and that $f'(x+h) > 0$ if $h > o$ (intuitively, f is increasing as we go right from x). Now, by the first derivative test, f has a local minimum at x.

Concavity Test

A related but distinct use of second derivatives is to determine whether a function is concave up or concave down at a point. It does not, however, provide information about inflection points. Specifically, a twice-differentiable function f is concave up if $f''(x) > 0$ and concave down if $f''(x) < 0$. Note that if $(x) = x^4$, then $x = 0$ has zero second derivative, yet is not an inflection point, so the second derivative alone does not give enough information to determine if a given point is an inflection point.

Higher-order Derivative Test

The *higher-order derivative test* or *general derivative test* is able to determine whether a function's critical points are maxima, minima, or points of inflection for a wider variety of functions than the second-order derivative test. The second derivative test is mathematically identical to the special case of n=1 in the higher-order derivative test.

Let f be a real-valued, sufficiently differentiable function on the interval $I \subset \mathbb{R}, c \in I$ and $n \geq 1$ an integer. Also let all the derivatives of f at c be zero up to and including the nth derivative, but with the $(n+1)$th derivative being non-zero:

$$f'(c) = \cdots = f^{(n)}(c) = 0 \quad \text{and} \quad f^{(n+1)}(c) \neq 0.$$

There are four possibilities, the first two cases where c is an extremum, the second two where c is a (local) saddle point:

- If n is odd and $f^{(n+1)}(c) < 0$, then c is a local maximum.

- If n is odd and $f^{(n+1)}(c) > 0$, then c is a local minimum.

- If n is even and $f^{(n+1)}(c) < 0$, then c is a strictly decreasing point of inflection.

- If n is even and $f^{(n+1)}(c) > 0$, then c is a strictly increasing point of inflection.

Since n must be either odd or even, this analytical test classifies any stationary point of f, so long as a nonzero derivative shows up eventually.

Example:

Say we want to perform the general derivative test on the function $f(x) = x^6 + 5$ at the point $x = 0$. To do this, we calculate the derivatives of the function and then evaluate them at the point of interest until the result is nonzero.

$$f'(x) = 6x^5, \; f'(0) = 0$$

$$f''(x) = 30x^4, \; f''(0) = 0$$

$$f^{(3)}(x) = 120x^3, \; f^{(3)}(0) = 0$$

$$f^{(4)}(x) = 360x^2, \; f^{(4)}(0) = 0$$

$$f^{(5)}(x) = 720x, \; f^{(5)}(0) = 0$$

$$f^{(6)}(x) = 720, \; f^{(6)}(0) = 720$$

At the point $x = 0$, the function $x^6 + 5$ has all of its derivatives at 0 equal to 0 except for the 6th derivative, which is positive. Thus n=5, and by the test, there is a local minimum at 0.

Multivariable Case

For a function of more than one variable, the second derivative test generalizes to a test based on the eigenvalues of the function's Hessian matrix at the critical point. In particular, assuming that all second order partial derivatives of f are continuous on a neighbourhood of a critical point x, then if the eigenvalues of the Hessian at x are all positive, then x is a local minimum. If the eigenvalues are all negative, then x is a local maximum, and if some are positive and some negative, then the point is a saddle point. If the Hessian matrix is singular, then the second derivative test is inconclusive.

Differentiation Rules

Differentiation rules is the rules for computing the derivative of a function in calculus.

Elementary Rules of Differentiation

Unless otherwise stated, all functions are functions of real numbers (R) that return real values; although more generally, the formulae below apply wherever they are well defined — including the case of complex numbers (C).

Differentiation is Linear

For any functions f and g and any real numbers a and b, the derivative of the function $h(x) = af(x) + bg(x)$ with respect to x is:

$$h'(x) = af'(x) + bg'(x)$$

In Leibniz's notation this is written as:

$$\frac{d(af + bg)}{dx} = a\frac{df}{dx} + b\frac{dg}{dx}.$$

Special cases include:

- The constant factor rule

    $$(af)' = af'$$

- The sum rule

    $$(f + g)' = f' + g'$$

- The subtraction rule

    $$(f - g)' = f' - g'.$$

The Product Rule

For the functions f and g, the derivative of the function $h(x) = f(x)\,g(x)$ with respect to x is

$$h'(x) = (fg)'(x) = f'(x)g(x) + f(x)g'(x).$$

In Leibniz's notation this is written

$$\frac{d(fg)}{dx} = \frac{df}{dx}g + f\frac{dg}{dx}.$$

The Chain Rule

The derivative of the function $h(x) = f(g(x))$ is:

$$h'(x) = f'(g(x)) \cdot g'(x).$$

In Leibniz's notation, this is written as:

$$\frac{d}{dx} h(x) = \frac{d}{dz} f(z)\big|_{z=g(x)} \cdot \frac{d}{dx} g(x),$$

often abridged to,

$$\frac{dh(x)}{dx} = \frac{df(g(x))}{dg(x)} \cdot \frac{dg(x)}{dx}.$$

Focusing on the notion of maps, and the differential being a map D, this is written in a more concise way as:

$$[D(h \circ g)]_x = [Dh]_{g(x)} \cdot [Dg]_x.$$

The Inverse Function Rule

If the function f has an inverse function g, meaning that $g(f(x)) = x$ and $f(g(y)) = y$, then

$$g' = \frac{1}{f' \circ g}.$$

In Leibniz notation, this is written as

$$\frac{dx}{dy} = \frac{1}{\dfrac{dy}{dx}}.$$

Power Laws, Polynomials, Quotients and Reciprocals

The Polynomial or Elementary Power Rule

If $f(x) = x^r$, for any real number $r \neq 0$ then

$$f'(x) = rx^{r-1}.$$

When $r = 1$, this becomes the special case that if $f(x) = x$, then $f'(x) = 1$.

Combining the power rule with the sum and constant multiple rules permits the computation of the derivative of any polynomial.

The Reciprocal Rule

The derivative of $h(x) = \dfrac{1}{f(x)}$ for any (nonvanishing) function f is:

$$h'(x) = -\frac{f'(x)}{(f(x))^2} \quad \text{wherever } f \text{ is non-zero.}$$

In Leibniz's notation, this is written

$$\frac{d(1/f)}{dx} = -\frac{1}{f^2}\frac{df}{dx}.$$

The reciprocal rule can be derived either from the quotient rule, or from the combination of power rule and chain rule.

The Quotient Rule

If f and g are functions, then:

$$\left(\frac{f}{g}\right)' = \frac{f'g - g'f}{g^2} \quad \text{wherever } g \text{ is nonzero.}$$

This can be derived from the product rule and the reciprocal rule.

Generalized Power Rule

The elementary power rule generalizes considerably. The most general power rule is the functional power rule: for any functions f and g,

$$(f^g)' = \left(e^{g \ln f}\right)' = f^g\left(f'\frac{g}{f} + g' \ln f\right),$$

wherever both sides are well defined.

Special cases

- If $f(x) = x^a$, then $f'(x) = ax^{a-1}$ when a is any non-zero real number and x is positive.

- The reciprocal rule may be derived as the special case where $g(x) = -1$.

Derivatives of Exponential and Logarithmic Functions

$$\frac{d}{dx}\left(c^{ax}\right) = ac^{ax} \ln c, \qquad c > 0$$

the equation above is true for all c, but the derivative for $c < 0$ yields a complex number.

$$\frac{d}{dx}\left(e^{ax}\right) = ae^{ax}$$

$$\frac{d}{dx}\left(\log_c x\right) = \frac{1}{x \ln c}, \qquad c > 0, c \neq 1$$

the equation above is also true for all c, but yields a complex number if $c < 0$.

$$\frac{d}{dx}\left(\ln x\right) = \frac{1}{x}, \qquad x > 0.$$

$$\frac{d}{dx}\left(\ln |x|\right) = \frac{1}{x}.$$

$$\frac{d}{dx}\left(x^x\right) = x^x(1 + \ln x).$$

$$\frac{d}{dx}\left(f(x)^{g(x)}\right) = g(x)f(x)^{g(x)-1}\frac{df}{dx} + f(x)^{g(x)}\ln(f(x))\frac{dg}{dx},$$

if $f(x) > 0$, and if $\dfrac{df}{dx}$ and $\dfrac{dg}{dx}$ exist

$$\frac{d}{dx}\left(f_1(x)^{f_2(x)^{(\ldots)^{f_n(x)}}}\right) = \left[\sum_{k=1}^{n}\frac{\partial}{\partial x_k}\left(f_1(x_1)^{f_2(x_2)^{(\ldots)^{f_n(x_n)}}}\right)\right]\Bigg|_{x_1 = x_2 = \ldots = x_n = x}, \text{ if } f_{i<n}(x) > 0 \text{ and } \frac{df_i}{dx} \text{ exists.}$$

Logarithmic Derivatives

The logarithmic derivative is another way of stating the rule for differentiating the logarithm of a function (using the chain rule):

$$(\ln f)' = \frac{f'}{f} \text{ wherever } f \text{ is positive.}$$

Logarithmic differentiation is a technique which uses logarithms and its differentiation rules to simplify certain expressions before actually applying the derivative. Logarithms can be used to remove exponents, convert products into sums, and convert division into subtraction — each of which may lead to a simplified expression for taking derivatives.

Derivatives of Trigonometric Functions

$(\sin x)' = \cos x$	$(\arcsin x)' = \dfrac{1}{\sqrt{1-x^2}}$
$(\cos x)' = -\sin x$	$(\arccos x)' = -\dfrac{1}{\sqrt{1-x^2}}$
$(\tan x)' = \sec^2 x = \dfrac{1}{\cos^2 x} = 1 + \tan^2 x$	$(\arctan x)' = \dfrac{1}{1+x^2}$
$(\sec x)' = \sec x \tan x$	$(\text{arcsec}\, x)' = \dfrac{1}{\lvert x \rvert \sqrt{x^2-1}}$
$(\csc x)' = -\csc x \cot x$	$(\text{arccsc}\, x)' = -\dfrac{1}{\lvert x \rvert \sqrt{x^2-1}}$
$(\cot x)' = -\csc^2 x = \dfrac{-1}{\sin^2 x} = -(1 + \cot^2 x)$	$(\text{arccot}\, x)' = -\dfrac{1}{1+x^2}$

It is common to additionally define an inverse tangent function with two arguments, $\arctan(y,x)$. Its value lies in the range $[-\pi, \pi]$ and reflects the quadrant of the point (x, y). For the first and fourth quadrant (i.e. $x > 0$) one has $\arctan(y, x > 0) = \arctan(y/x)$. Its partial derivatives are:

$$\frac{\partial \arctan(y,x)}{\partial y} = \frac{x}{x^2 + y^2}, \text{ and } \frac{\partial \arctan(y,x)}{\partial x} = \frac{-y}{x^2 + y^2}.$$

Derivatives of Hyperbolic Functions

$(\sinh x)' = \cosh x = \dfrac{e^x + e^{-x}}{2}$	$(\text{arsinh}\, x)' = \dfrac{1}{\sqrt{x^2 + 1}}$
$(\cosh x)' = \sinh x = \dfrac{e^x - e^{-x}}{2}$	$(\cosh x)' = \sinh x = \dfrac{e^x - e^{-x}}{2}$
$(\tanh x)' = \text{sech}^2 x$	$(\text{artanh}\, x)' = \dfrac{1}{1-x^2}$
$(\text{sech}\, x)' = -\tanh x \,\text{sech}\, x$	$(\text{arsech}\, x)' = -\dfrac{1}{x\sqrt{1-x^2}}$
$(\text{csch}\, x)' = -\coth x \,\text{csch}\, x$	$(\text{arcsch}\, x)' = -\dfrac{1}{\lvert x \rvert \sqrt{1+x^2}}$
$(\coth x)' = -\text{csch}^2 x$	$(\text{arcoth}\, x)' = \dfrac{1}{1-x^2}$

Derivatives of Special Functions

Gamma function $\Gamma(x) = \int_0^\infty t^{x-1} e^{-t}\, dt$

$$\Gamma'(x) = \int_0^\infty t^{x-1} e^{-t} \ln t\, dt$$

$$= \Gamma(x) \left(\sum_{n=1}^\infty \left(\ln\left(1+\frac{1}{n}\right) - \frac{1}{x+n} \right) - \frac{1}{x} \right)$$

$$= \Gamma(x)\psi(x)$$

with $\psi(x)$ being the digamma function, expressed by the parenthesized expression to the right of $\tilde{A}(x)$ in the line above.

Riemann Zeta function $\zeta(x) = \sum_{n=1}^\infty \frac{1}{n^x}$

$$\zeta'(x) = -\sum_{n=1}^\infty \frac{\ln n}{n^x} = -\frac{\ln 2}{2^x} - \frac{\ln 3}{3^x} - \frac{\ln 4}{4^x} - \cdots$$

$$= -\sum_{p \text{ prime}} \frac{p^{-x} \ln p}{(1-p^{-x})^2} \prod_{q \text{ prime}, q \neq p} \frac{1}{1-q^{-x}}$$

Derivatives of Integrals

Suppose that it is required to differentiate with respect to x the function:

$$F(x) = \int_{a(x)}^{b(x)} f(x,t)\, dt,$$

where the functions $f(x,t)$ and $\frac{\partial}{\partial x} f(x,t)$ are both continuous in both t and x in some region of the (t,x) plane, including $a(x) \leq t \leq b(x)$ $x_0 \leq x \leq x_1$, and the functions $a(x)$ and $b(x)$ are both continuous and both have continuous derivatives for $x_0 \leq x \leq x_1$. Then for $x_0 \leq x \leq x_1$:

$$F'(x) = f(x,b(x))b'(x) - f(x,a(x))a'(x) + \int_{a(x)}^{b(x)} \frac{\partial}{\partial x} f(x,t)\, dt.$$

This formula is the general form of the Leibniz integral rule and can be derived using the fundamental theorem of calculus.

Derivatives to nth Order

Some rules exist for computing the n-th derivative of functions, where n is a positive integer. These include.

Faà di Bruno's Formula

If f and g are n-times differentiable, then

$$\frac{d_n}{dx^n}[f(g(x))] = n! \sum_{\{k_m\}} f^{(r)}(g(x)) \prod_{m=1}^{n} \frac{1}{k_m!} \left(g^{(m)}(x) \right)^{k_m}$$

where $r = \sum_{m=1}^{n-1} k_m$ and the set $\{k_m\}$ consists of all non-negative integer solutions of the Diophantine equation $\sum_{m=1}^{n} m k_m = n$.

General Leibniz Rule

If f and g are n-times differentiable, then

$$\frac{d^n}{dx^n}[f(x)g(x)] = \sum_{k=0}^{n} \binom{n}{k} \frac{d^{n-k}}{dx^{n-k}} f(x) \frac{d^k}{dx^k} g(x).$$

Differential Equation

A differential equation is a mathematical equation that relates some function with its derivatives. In applications, the functions usually represent physical quantities, the derivatives represent their rates of change, and the differential equation defines a relationship between the two. Because such relations are extremely common, differential equations play a prominent role in many disciplines including engineering, physics, economics, and biology.

In pure mathematics, differential equations are studied from several different perspectives, mostly concerned with their solutions—the set of functions that satisfy the equation. Only the simplest differential equations are solvable by explicit formulas; however, some properties of solutions of a given differential equation may be determined without finding their exact form.

If a closed-form expression for the solution is not available, the solution may be numerically approximated using computers. The theory of dynamical systems puts emphasis on qualitative analysis of systems described by differential equations, while many numerical methods have been developed to determine solutions with a given degree of accuracy.

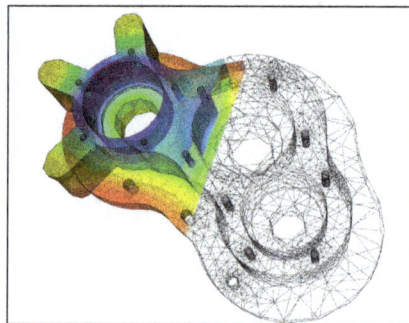

Visualization of heat transfer in a pump casing, created by solving the heat equation. Heat is being generatedinternally in the casing and being cooled at the boundary, providing a steady state temperature distribution.

For example, in classical mechanics, the motion of a body is described by its position and velocity as the time value varies. Newton's laws allow these variables to be expressed dynamically (given the position, velocity, acceleration and various forces acting on the body) as a differential equation for the unknown position of the body as a function of time.

In some cases, this differential equation (called an equation of motion) may be solved explicitly. An example of modeling a real-world problem using differential equations is the determination of the velocity of a ball falling through the air, considering only gravity and air resistance. The ball's acceleration towards the ground is the acceleration due to gravity minus the acceleration due to air resistance. Gravity is considered constant, and air resistance may be modeled as proportional to the ball's velocity. This means that the ball's acceleration, which is a derivative of its velocity, depends on the velocity (and the velocity depends on time). Finding the velocity as a function of time involves solving a differential equation and verifying its validity.

Types

Differential equations can be divided into several types. Apart from describing the properties of the equation itself, these classes of differential equations can help inform the choice of approach to a solution. Commonly used distinctions include whether the equation is: Ordinary/Partial, Linear/Non-linear, and Homogeneous/heterogeneous. This list is far from exhaustive; there are many other properties and subclasses of differential equations which can be very useful in specific contexts.

Ordinary Differential Equations

An ordinary differential equation (*ODE*) is an equation containing an unknown function of one real or complex variable x, its derivatives, and some given functions of x. The unknown function is generally represented by a variable (often denoted y), which, therefore, *depends* on x. Thus x is often called the independent variable of the equation. The term "*ordinary*" is used in contrast with the term partial differential equation, which may be with respect to *more than* one independent variable.

Linear differential equations are the differential equations that are linear in the unknown function and its derivatives. Their theory is well developed, and, in many cases, one may express their solutions in terms of integrals.

Most ODEs that are encountered in physics are linear, and, therefore, most special functions may be defined as solutions of linear differential equations. As, in general, the solutions of a differential equation cannot be expressed by a closed-form expression, numerical methods are commonly used for solving differential equations on a computer.

Partial Differential Equations

A partial differential equation (*PDE*) is a differential equation that contains unknown multivariable functions and their partial derivatives. (This is in contrast to ordinary differential equations, which deal with functions of a single variable and their derivatives.) PDEs are used to formulate problems involving functions of several variables, and are either solved in closed form, or used to create a relevant computer model.

PDEs can be used to describe a wide variety of phenomena in nature such as sound, heat, electrostatics, electrodynamics, fluid flow, elasticity, or quantum mechanics. These seemingly distinct physical phenomena can be formalized similarly in terms of PDEs. Just as ordinary differential equations often model one-dimensional dynamical systems, partial differential equations often model multidimensional systems. PDEs find their generalization in stochastic partial differential equations.

Non-linear Differential Equations

A non-linear differential equation is a differential equation that is not a linear equation in the unknown function and its derivatives (the linearity or non-linearity in the arguments of the function are not considered here). There are very few methods of solving nonlinear differential equations exactly; those that are known typically depend on the equation having particular symmetries. Nonlinear differential equations can exhibit very complicated behavior over extended time intervals, characteristic of chaos. Even the fundamental questions of existence, uniqueness, and extendability of solutions for nonlinear differential equations, and well-posedness of initial and boundary value problems for nonlinear PDEs are hard problems and their resolution in special cases is considered to be a significant advance in the mathematical theory. However, if the differential equation is a correctly formulated representation of a meaningful physical process, then one expects it to have a solution.

Linear differential equations frequently appear as approximations to nonlinear equations. These approximations are only valid under restricted conditions. For example, the harmonic oscillator equation is an approximation to the nonlinear pendulum equation that is valid for small amplitude oscillations.

Equation Order

Differential equations are described by their order, determined by the term with the highest derivatives. An equation containing only first derivatives is a *first-order differential equation*, an equation containing the second derivative is a *second-order differential equation*, and so on. Differential equations that describe natural phenomena almost always have only first and second order derivatives in them, but there are some exceptions, such as the thin film equation, which is a fourth order partial differential equation.

Examples:

In the first group of examples, u is an unknown function of x, and c and ω are constants that are supposed to be known. Two broad classifications of both ordinary and partial differential equations consists of distinguishing between *linear* and *nonlinear* differential equations, and between *homogeneous* differential equations and *heterogeneous* ones.

- Heterogeneous first-order linear constant coefficient ordinary differential equation:

$$\frac{du}{dx} = cu + x^2.$$

- Homogeneous second-order linear ordinary differential equation:

$$\frac{d^2u}{dx^2} - x\frac{du}{dx} + u = 0.$$

- Homogeneous second-order linear constant coefficient ordinary differential equation describing the harmonic oscillator:

$$\frac{d^2u}{dx^2} + \omega^2 u = 0.$$

- Heterogeneous first-order nonlinear ordinary differential equation:

$$\frac{du}{dx} = u^2 + 4.$$

- Second-order nonlinear (due to sine function) ordinary differential equation describing the motion of a pendulum of length L:

$$L\frac{d^2u}{dx^2} + g\sin u = 0.$$

In the next group of examples, the unknown function u depends on two variables x and t or x and y.

- Homogeneous first-order linear partial differential equation:

$$\frac{\partial u}{\partial t} + t\frac{\partial u}{\partial x} = 0.$$

- Homogeneous second-order linear constant coefficient partial differential equation of elliptic type, the Laplace equation:

$$\frac{\partial^2 u}{\partial x^2} + \frac{\partial^2 u}{\partial y^2} = 0.$$

- Homogeneous third-order non-linear partial differential equation:

$$\frac{\partial u}{\partial t} = 6u\frac{\partial u}{\partial x} - \frac{\partial^3 u}{\partial x^3}.$$

Existence of Solutions

Solving differential equations is not like solving algebraic equations. Not only are their solutions often unclear, but whether solutions are unique or exist at all are also notable subjects of interest.

For first order initial value problems, the Peano existence theorem gives one set of circumstances in which a solution exists. Given any point (a,b) in the xy-plane, define some rectangular region Z, such that $Z = [l,m] \times [n,p]$ and (a,b) is in the interior of Z. If we are given a differential equation $\dfrac{dy}{dx} = g(x, y)$ and the condition that $y = bw$ when $x = a$, then there is locally a solution to this problem if $g(x, y)$ and $\dfrac{\partial g}{\partial x}$ are both continuous on Z. This solution exists on some interval with its center at a. The solution may not be unique.

However, this only helps us with first order initial value problems. Suppose we had a linear initial value problem of the nth order:

$$f_n(x)\frac{d^n y}{dx^n} + \cdots + f_1(x)\frac{dy}{dx} + f_0(x)y = g(x)$$

such that,

$$y(x_0) = y_0, y'(x_0) = y'_0, y''(x_0) = y''_0$$

For any nonzero $f_n(x)$, if $\{f_0, f_1, \cdots\}$ and g are continuous on some interval containing x_0, y is unique and exists.

Applications

The study of differential equations is a wide field in pure and applied mathematics, physics, and engineering. All of these disciplines are concerned with the properties of differential equations of various types. Pure mathematics focuses on the existence and uniqueness of solutions, while applied mathematics emphasizes the rigorous justification of the methods for approximating solutions. Differential equations play an important role in modeling virtually every physical, technical, or biological process, from celestial motion, to bridge design, to interactions between neurons. Differential equations such as those used to solve real-life problems may not necessarily be directly solvable, i.e. do not have closed form solutions. Instead, solutions can be approximated using numerical methods.

Many fundamental laws of physics and chemistry can be formulated as differential equations. In biology and economics, differential equations are used to model the behavior of complex systems. The mathematical theory of differential equations first developed together with the sciences where the equations had originated and where the results found application. However, diverse problems, sometimes originating in quite distinct scientific fields, may give rise to identical differential equations. Whenever this happens, mathematical theory behind the equations can be viewed as a unifying principle behind diverse phenomena. As an example, consider the propagation of light and sound in the atmosphere, and of waves on the surface of a pond. All of them may be described by the same second-order partial differential equation, the wave equation, which allows us to think of light and sound as forms of waves, much like familiar waves in the water. Conduction of heat, the theory of which was developed by Joseph Fourier, is governed by another second-order partial differential equation, the heat equation. It turns out that many diffusion processes, while seemingly different, are described by the same equation; the Black–Scholes equation in finance is, for instance, related to the heat equation.

INTEGRAL CALCULUS

The branch of mathematics in which the notion of an integral, its properties and methods of calculation are studied. Integral calculus is intimately related to differential calculus, and together with it constitutes the foundation of mathematical analysis. The origin of integral calculus goes back to the early period of development of mathematics and it is related to the method of exhaustion developed by the mathematicians of Ancient Greece (cf. Exhaustion, method of). This method arose in the solution of problems on calculating areas of plane figures and surfaces, volumes of solid bodies, and in the solution of certain problems in statistics and hydrodynamics. It is based on the approximation of the objects under consideration by stepped figures or bodies, composed of simplest planar figures or special bodies (rectangles, parallelopipeds, cylinders, etc.). In this sense, the method of exhaustion can be regarded as an early method of integration. The greatest development of the method of exhaustion in the early period was obtained in the works of Eudoxus (4th century B.C.) and especially Archimedes (3rd century B.C.). Its subsequent application and perfection is associated with the names of several scholars of the 15th–17th centuries.

The fundamental concepts and theory of integral and differential calculus, primarily the relationship between differentiation and integration, as well as their application to the solution of applied problems, were developed in the works of P. de Fermat, I. Newton and G. Leibniz at the end of the 17th century. Their investigations were the beginning of an intensive development of mathematical analysis. The works of L. Euler, Jacob and Johann Bernoulli and J.L. Lagrange played an essential role in its creation in the 18th century. In the 19th century, in connection with the appearance of the notion of a limit, integral calculus achieved a logically complete form (in the works of A.L. Cauchy, B. Riemann and others). The development of the theory and methods of integral calculus took place at the end of 19th century and in the 20th century simultaneously with research into measure theory (cf. Measure), which plays an essential role in integral calculus.

By means of integral calculus it became possible to solve by a unified method many theoretical and applied problems, both new ones which earlier had not been amenable to solution, and old ones that had previously required special artificial techniques. The basic notions of integral calculus are two closely related notions of the integral, namely the indefinite and the definite integral.

The indefinite integral of a given real-valued function on an interval on the real axis is defined as the collection of all its primitives on that interval, that is, functions whose derivatives are the given function. The indefinite integral of a function f is denoted by $\int f(x)dx$. If F is some primitive of f, then any other primitive of it has the form $F + C$, where C is an arbitrary constant; one therefore writes

$$\int f(x)dx = F(x) + C$$

The operation of finding an indefinite integral is called integration. Integration is the operation inverse to that of differentiation:

$$\int f'(x)dx = F(x) + C, \quad d\int f(x)dx = f(x)\,dx .$$

The operation of integration is linear: If on some interval the indefinite integrals exist, then for any real numbers λ_1 and λ_2, the following integral exists on this interval:

$$\int f_1(x)dx \quad \text{and} \quad \int f_2(x)dx$$

$$\int [\lambda_1 f_1(x) + \lambda_2 f_2(x)]dx$$

and equals,

$$\lambda_1 \int f_1(x)dx + \lambda_2 \int f_2(x)dx.$$

For indefinite integrals, the formula of integration by parts holds: If two functions u and v are differentiable on some interval and if the integral $\int vdu$ exists, then so does the integral $\int udu$, and the following formula holds:

$$\int udv = uv - \int vdu$$

The formula for change of variables holds: If for two functions f and ϕ defined on certain intervals, the composite function $f \circ \phi$ makes sense and the function ϕ is differentiable, then the integral,

$$\int f[\phi(t)]\phi'(t)dt$$

exists and equals,

$$\int f(x)dx$$

A function that is continuous on some bounded interval has a primitive on it and hence an indefinite integral exists for it. The problem of actually finding the indefinite integral of a specified function is complicated by the fact that the indefinite integral of an elementary function is not an elementary function, in general. Many classes of functions are known for which it proves possible to express their indefinite integrals in terms of elementary functions. The simplest examples of these are integrals that are obtained from a table of derivatives of the basic elementary functions:

1) $\int x^{\alpha}dx = \dfrac{x^{\alpha+1}}{\alpha+1} + C, \alpha \neq -1;$

2) $\int \dfrac{dx}{x} = \ln|x| + C;$

3) $\int \alpha^x dx = \dfrac{\alpha^x}{\ln \alpha} + C, \ \alpha > 0, \ \alpha \neq 1;$ in particular, $\int e^x dx = e^x + C;$

4) $\int \sin x \, dx = -\cos x + C;$

5) $\int \cos x \, dx = \sin x + C;$

6) $\int \dfrac{dx}{\cos^2 x} = \tan x + C;$

7) $\int \dfrac{dx}{\sin^2 x} = -\cotan x + C;$

8) $\int \sinh x\, dx = \cosh x + C;$

9) $\int \cosh x\, dx = \sinh x + C;$

10) $\int \dfrac{dx}{\cosh^2 x} = \tanh x + C;$

11) $\int \dfrac{dx}{\sinh^2 x} = \cotanh x + C;$

12) $\int \dfrac{dx}{x^2 + \alpha^2} = \dfrac{1}{\alpha} \arctan \dfrac{x}{\alpha} + C = -\dfrac{1}{\alpha} \arccotan \dfrac{x}{\alpha} + C';$

13) $\int \dfrac{dx}{x^2 - \alpha^2} = \dfrac{1}{2\alpha} \ln \left| \dfrac{x-\alpha}{x+\alpha} \right| + C;$

14) $\int \dfrac{dx}{\sqrt{\alpha^2 - x^2}} = \arcsin \dfrac{x}{\alpha} + C = -\arccos \dfrac{x}{\alpha} + C', |x| < |\alpha|;$

15) $\int \dfrac{dx}{\sqrt{x^2 \pm \alpha^2}} = \ln |x + \sqrt{x^2 \pm \alpha^2}| + C$ (when $x^2 - \alpha^2$ is under the square root, it is assumed that $|x| > |\alpha|$).

If the denominator of the integrand vanishes at some point, then these formulas are valid only for those intervals inside which the denominator does not vanish.

The indefinite integral of a rational function over any interval on which the denominator does not vanish is a composition of rational functions, arctangents and natural logarithms. Finding the algebraic part of the indefinite integral of a rational function can be achieved by the Ostrogradski method. Integrals of the following types can be reduced by means of substitution and integration by parts to integration of rational functions:

$$\int R\left[x, \left(\frac{\alpha x + b}{cx + b} \right)^{r_1}, ..., \left(\frac{\alpha x + b}{cx + b} \right)^{r_m} \right] dx,$$

where $r_1, .., r_m$ are rational numbers; integrals of the form,

$$\int R(x, \sqrt{ax^2 + bx + c})\, dx$$

Certain cases of integrals of differential binomials; integrals of the form,

$$\int R(\sin x, \cos x)\, dx, \quad \int R(\sinh x, \cosh x)\, dx$$

(where $R(y_1, ..., y_n)$ are rational functions); the integrals,

$$\int e^{\alpha x} \cos \beta x\, dx, \quad \int e^{\alpha x} \sin \beta x\, dx,$$

$$\int x^n \cos \alpha x\, dx, \quad \int x^n \sin \alpha x\, dx,$$

$$\int x^n \arcsin x \, dx, \quad \int x^n \arccos x \, dx,$$

$$\int x^n \arctan x \, dx, \quad \int x^n \operatorname{arccotan} x \, dx, \quad n \quad 0, 1, \ldots,$$

and many others. In contrast, for example, the integrals,

$$\int \frac{e^x}{x^n} \, dx, \quad \int \frac{\sin x}{x^n} \, dx, \quad \int \frac{\cos x}{x^n} \, dx, \quad n = 1, 2, \ldots,$$

cannot be expressed in terms of elementary functions.

The definite integral,

$$\int_a^b f(x) \, dx$$

of a function f defined on an interval $[a, b]$ is the limit of integral sums of a specific type. If this limit exists, f is said to be Cauchy, Riemann, Lebesgue, etc. integrable.

The geometrical meaning of the integral is tied up with the notion of area: If the function $f \geq 0$ is continuous on the interval $[a, b]$, then the value of the integral,

$$\int_a^b f(x) \, dx$$

is equal to the area of the curvilinear trapezium formed by the graph of the function, that is, the set whose boundary consists of the graph of f, the segment $[a, b]$ and the two segments on the lines $x = a$ and $x = b$ making the figure closed, which may degenerate to points.

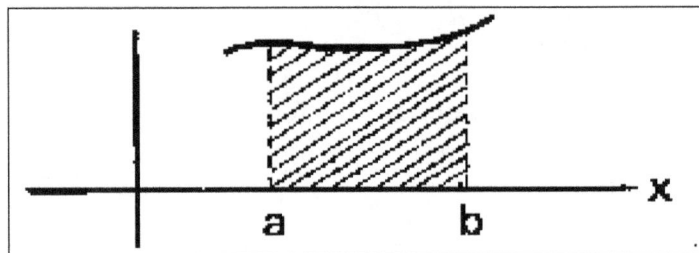

The calculation of many quantities encountered in practice reduces to the problem of calculating the limit of integral sums; in other words, finding a definite integral; for example, areas of figures and surfaces, volumes of bodies, work done by force, the coordinates of the centre of gravity, the values of the moments of inertia of various bodies, etc.

The definite integral is linear: If two functions f_1 and f_2 are integrable on an interval, then for any real numbers λ_1 and λ_2 the function:

$$\lambda_1 f_1 + \lambda_2 f_2$$

is also integrable on this interval and

$$\int_\alpha^b \left[\lambda_1 f_1(x) + \lambda_2 f_2(x)\right] dx = \lambda_1 \int_\alpha^b f_1(x)dx + \lambda_2 \int_\alpha^b f_2(x)\, dx.$$

Integration of a function over an interval has the property of monotonicity: If the function f is integrable on the interval $[a, b]$ and if $[c, d] \subset [a, b]$, then f is integrable on $[c, d]$ as well. The integral is also additive with respect to the intervals over which the integration is carried out: If $a < c < b$ and the function f is integrable on the intervals $[a, c]$ and $[c, d]$, then it is integrable on $[a, b]$, and

$$\int_\alpha^b f(x)\, dx = \int_\alpha^c f(x)\, dx + \int_c^b f(x)\, dx.$$

If f and g are Riemann integrable, then their product is also Riemann integrable. If $f \geq g$ on $[a, b]$ then,

$$\int_\alpha^b f(x)\, dx \geq \int_\alpha^b g(x)\, dx.$$

If f is integrable on $[a, b]$, then the absolute value $|f|$ is also integrable on $[a, b]$ if $-\infty < a < b\, \infty$, and

$$\left| \int_\alpha^b f(x)\, dx \right| \leq \int_\alpha^b |f(x)|\, dx.$$

By definition one sets,

$$\int_\alpha^\alpha f(x)dx = 0 \ \text{ and } \ \int_b^\alpha f(x)dx = -\int_\alpha^b f(x)dx, \qquad a < b.$$

A mean-value theorem holds for integrals. For example, if f and g are Riemann integrable on an interval $[a, b]$, if $m \leq f(x) \leq M$, $x \in [a, b]$, and if g does not change sign on $[a, b]$, that is, it is either non-negative or non-positive throughout this interval, then there exists a number $m \leq \mu \leq M$ for which,

$$\int_\alpha^b f(x)g(x)\, dx = \mu \int_\alpha^b g(x)\, dx.$$

Under the additional hypothesis that f is continuous on $[a, b]$, there exists in (a, b) a point ξ for which,

$$\int_\alpha^b f(x)g(x)\, dx = f(\xi) \int_\alpha^b g(x)\, dx.$$

In particular, if $g(x) \equiv 1$, then

$$\int_\alpha^b f(x)dx = f(\xi)(b - \alpha).$$

Integrals with a Variable upper Limit

If a function f is Riemann integrable on an interval $[a, b]$, then the function F defined by:

$$F(x) = \int_\alpha^x f(t)\,dt, \quad a \le x \le b,$$

is continuous on this interval. If, in addition, f is continuous at a point x_0, then F is differentiable at this point and $F'(x_0) = f(x_0)$. In other words, at the points of continuity of a function the following formula holds:

$$\frac{d}{dx} \int_\alpha^x f(t)\,dt = f(x).$$

Consequently, this formula holds for every Riemann-integrable function on an interval $[a, b]$, except perhaps at a set of points having Lebesgue measure zero, since if a function is Riemann integrable on some interval, then its set of points of discontinuity has measure zero. Thus, if the function f is continuous on $[a, b]$, then the function F defined by,

$$F(x) = \int_\alpha^x f(t)\,dt$$

is a primitive of f on this interval. This theorem shows that the operation of differentiation is inverse to that of taking the definite integral with a variable upper limit, and in this way a relationship is established between definite and indefinite integrals:

$$\int f(x)\,dx = \int_\alpha^x f(t)\,dt + C$$

The geometric meaning of this relationship is that the problem of finding the tangent to a curve and the calculation of the area of plane figures are inverse operations in the above sense.

The following Newton–Leibniz formula holds for any primitive F of an integrable function f on an interval $[a, b]$:

$$\int_\alpha^b f(x)\,dx = F(b) - F(a)$$

It shows that the definite integral of a continuous function over some interval is equal to the difference of the values at the end points of this interval of any primitive of it. This formula is sometimes taken as the definition of the definite integral. Then it is proved that the integral $\int_a^b f(x)\,dx$ introduced in this way is equal to the limit of the corresponding integral sums.

For definite integrals, the formulas for change of variables and integration by parts hold. Suppose, for example, that the function f is continuous on the interval (a, b) and that ϕ is continuous together with its derivative ϕ' on the interval (α, β), where (α, β) is mapped by ϕ into (a, b): $a < \phi(t) < b$ for $\alpha < t < \beta$, so that the composite $f \circ \phi$ is meaningful in (α, β). Then, for $\alpha_0, \beta_0 \in (\alpha, \beta)$, the following formulas for change of variables holds:

$$\int_{\phi(\alpha_0)}^{\phi(\beta_0)} f(x)\,dx = \int_{\alpha_0}^{\beta_0} f[\phi(t)]\,\phi'(t)\,dt.$$

The formula for integration by parts is:

$$\int_a^b u(x)\,dv(x) = u(x)v(x)\,\big|_{x=a}^{x=b} - \int_a^b v(x)\,du(x)$$

where the functions u and v have Riemann-integrable derivatives on $[a, b]$.

The Newton–Leibniz formula reduces the calculation of an indefinite integral to finding the values of its primitive. Since the problem of finding a primitive is intrinsically a difficult one, other methods of finding definite integrals are of great importance, among which one should mention the method of residues and the method of differentiation or integration with respect to the parameter of a parameter-dependent integral. Numerical methods for the approximate computation of integrals have also been developed.

Generalizing the notion of an integral to the case of unbounded functions and to the case of an unbounded interval leads to the notion of the improper integral, which is defined by yet one more limit transition. The notions of the indefinite and the definite integral carry over to complex-valued functions. The representation of any holomorphic function of a complex variable in the form of a Cauchy integral over a contour played an important role in the development of the theory of analytic functions.

The generalization of the notion of the definite integral of a function of a single variable to the case of a function of several variables leads to the notion of a multiple integral. For unbounded sets and unbounded functions of several variables, one is led to the notion of the improper integral, as in the one-dimensional case.

The extension of the practical applications of integral calculus necessitated the introduction of the notions of the curvilinear integral, i.e. the integral along a curve, the surface integral, i.e. the integral over a surface, and more generally, the integral over a manifold, which are reducible in some sense to a definite integral (the curvilinear integral reduces to an integral over an interval, the surface integral to an integral over a (plane) region, the integral over an n-dimensional manifold to an integral over an n-dimensional region). Integrals over manifolds, in particular curvilinear and surface integrals, play an important role in the integral calculus of functions of several variables; by this means a relationship is established between integration over a region and integration over its boundary or, in the general case, over a manifold and its boundary. This relationship is established by the Stokes formula (which is a generalization of the Newton–Leibniz formula to the multi-dimensional case.

Multiple, curvilinear and surface integrals find direct application in mathematical physics, particularly in field theory. Multiple integrals and concepts related to them are widely used in the solution of specific applied problems. The theory of cubature formulas has been developed for the numerical calculation of multiple integrals.

The theory and methods of integral calculus of real- or complex-valued functions of a finite number of real or complex variables carry over to more general objects. For example, the theory of integration of functions whose values lie in a normed linear space, functions defined on topological groups, generalized functions, and functions of an infinite number of variables (integrals over trajectories). Finally, a new direction in integral calculus is related to the emergence and development of constructive mathematics.

Integral calculus is applied in many branches of mathematics (in the theory of differential and integral equations, in probability theory and mathematical statistics, in the theory of optimal processes, etc.), and in applications of it.

Integral

One of the central notions in mathematical analysis and all of mathematics, which arose in connection with two problems: to recover a function from its derivative (for example, the problem of finding the law of motion of a material object along a straight line when the velocity of this point is known); and to calculate the area bounded by the graph of a function f on an interval $a \leq x \leq b$ and the xx-axis (the problem of calculating the work performed by a force over an interval of time $a \leq t \leq b$ leads to this problem, as do other problems).

The two problems indicated above lead to two forms of the integral, the indefinite and the definite integral. The study of the properties and calculation of these interrelated forms of the integral constitutes the problem of integral calculus.

In the course of development of mathematics and under the influence of the requirements of natural science and technology, the notions of the indefinite and the definite integral have undergone a number of generalizations and modifications.

The Indefinite Integral

A primitive of a function   of the variable x on an interval $a < x < b$ is any function F whose derivative is equal to f at each point x of the interval. It is clear that if F is a primitive of f on the interval $a < x < b$ then so is $F_1 = F + C$, where C is an arbitrary constant. The converse also holds: Any two primitives of the same function f on the interval $a < x < b$ can only differ by a constant. Consequently, if F is one of the primitives of f on the interval $a < x < b$, then any primitive of f on this interval has the form $F + C$, where C is a constant. The collection of all primitives of f on the interval a<x<b is called the indefinite integral of f (on this interval) and is denoted by the symbol.

$$\int f(x)dx.$$

According to the fundamental theorem of integral calculus, there exists for each continuous function f on the interval $a < x < b$ a primitive, and hence an indefinite integral, on this interval.

The Definite Integral

The notion of the definite integral is introduced either as a limit of integral sums or, in the case when the given function f is defined on some interval $[a,b]$ and has a primitive F on this interval, as the difference between the values at the end points, that is, as $F(b) - F(a)$. The definite integral of f on $[a,b]$ is denoted by $\int_a^b f(x)dx$. The definition of the integral as a limit of integral sums for the case of continuous functions was stated by A.L. Cauchy in 1823. The case of arbitrary functions was studied by B. Riemann. A substantial advance in the theory of definite integrals was made by G. Darboux, who introduced the notion of upper and lower Riemann sums. A necessary and sufficient condition for the Riemann integrability of discontinuous functions was established in final form in 1902 by H. Lebesgue.

There is the following relationship between the definitions of the definite integral of a continuous function f on a closed interval [a,b] and the indefinite integral (or primitive) of this function: 1) if F is any primitive of f, then the following Newton–Leibniz formula holds:

$$\int_a^b f(x)dx = F(b) - F(a);$$

For any x in the interval $[a,b]$, the indefinite integral of the continuous function f can be written in the form

$$\int_a^b f(x)dx = \int_a^x f(t)dt + C$$

where C is an arbitrary constant. In particular, the definite integral with variable upper limit,

$$F(x) = \int_a^x f(t)dt$$

is a primitive of f.

In order to introduce the definite integral of f over $[a,b]$ in the sense of Lebesgue, the set of values of y is divided into subintervals of points $\ldots < y_{-1} < y_0 < y_1 < \ldots$, and one denotes by M_i the set of all values of x in the interval $[a,b]$ for which $y_{i-1} \le f(x) < y_i$, and by $\mu(M_i)$ the measure of the set M_i in the sense of Lebesgue. A Lebesgue integral sum of the function f on the interval $[a,b]$ is defined by the formula,

$$\sigma = \sum_i \eta_i \mu(M_i)$$

where η_i are arbitrary numbers in the interval $[y_{i-1}, y_i]$.

A function f is said to be Lebesgue integrable on the interval $[a, b]$ if the limit of the integral sums ??? exists and is finite as the maximum width of the intervals (y_{i-1}, y_i) tends to zero, that is, if there exists a real number I such that for any $\epsilon > 0$ there is a $\delta > 0$ such that under the single condition $\max(y_i - y_{i-1}) < \delta$ the inequality $|\sigma - I| < \epsilon$ holds. The limit I is then called the definite Lebesgue integral of f over $[a, b]$.

Instead of the interval $[a, b]$ one can consider an arbitrary set that is measurable with respect to some non-negative complete countably-additive measure. An alternative introduction to the Lebesgue integral can be given, when one defines this integral originally on the set of so-called simple functions (that is, measurable functions assuming at most a countable number of values), and then introduces the integral by means of a limit transition for any function that can be expressed as the limit of a uniformly-convergent sequence of simple functions.

Each Riemann-integrable function is Lebesgue integrable. The converse is false, since there exist Lebesgue-integrable functions that are discontinuous on a set of positive measure (for example, the Dirichlet function).

In order that a bounded function be Lebesgue integrable, it is necessary and sufficient that this function belongs to the class of measurable functions. The functions encountered in mathematical analysis are, as a rule, measurable. This means that the Lebesgue integral has a generality that is sufficient for the requirements of analysis.

The Lebesgue integral also covers the cases of absolutely-convergent improper integrals.

The generality attained by the definition of the Lebesgue integral is absolutely essential in many questions in modern mathematical analysis (the theory of generalized functions, the definition of generalized solutions of differential equations, and the isomorphism of the Hilbert spaces L_2 and l_2, which is equivalent to the so-called Riesz–Fischer theorem in the theory of trigonometric or arbitrary orthogonal series; all these theories have proved possible only by taking the integral to be in the sense of Lebesgue).

The primitive in the sense of Lebesgue is naturally defined by means of equation ???, in which the integral is taken in the sense of Lebesgue. The relation $F' = f$ in this case holds everywhere, except perhaps on a set of measure zero.

Other Generalizations of the Notions of an Integral

In 1894 T.J. Stieltjes gave another generalization of the Riemann integral (which acquired the name of Stieltjes integral), important for applications, in which one considers the integrability of a function f defined on some interval $[a, b]$ with respect to a second function defined on the same interval. The Stieltjes integral of f with respect to the function U is denoted by the symbol,

$$I = \int_a^b f(x) dU(x).$$

If U has a bounded Riemann-integrable derivative U', then the Stieltjes integral reduces to the Riemann integral by the formula:

$$\int_a^b f(x)dU(x) = \int_a^b f(x)U'(x)dx.$$

In particular, when $U(x) = x + C$, the Stieltjes integral ??? is the Riemann integral $\int_a^b f(x)dx$.

However, the interesting case for applications is when the function U does not have a derivative. An example of such a U is the spectral measure in the study of spectral decompositions.

The curvilinear integral,

$$\int_{\tilde{A}} f(x,y)dx$$

along the curve $\tilde{A}$ defined by the equations $x = \phi(t)$, $y = \psi(t)$, y=ψ(t), $a \le t \le b$, is a special case of the Stieltjes integral, since it can be written in the form,

$$\int_a^b f[\phi(t), \psi(t)]d\phi(t).$$

A further generalization of the notion of the integral is obtained by integration over an arbitrary set in a space of any number of variables. In the most general case it is convenient to regard the integral as a function of the set M over which the integration is carried out, in the form,

$$F(M) = \int_M f(x)dU(x),$$

where U is a set function on M (its measure in a particular case) and the points belong to the set M over which the integration proceeds. Particular cases of this type of integration are multiple integrals and surface integrals.

Another generalization of the notion of the integral is that of the improper integral.

In 1912 A. Denjoy introduced a notion of the integral that can be applied to every function f that is the derivative of some function F. This enables one to reduce the constructive definition of the integral to a degree of generality which completely answers the problem of finding a definite integral taken in the sense of a primitive.

Quadratic Integral

In mathematics, a quadratic integral is an integral of the form,

$$\int \frac{dx}{a + bx + cx^2},$$

It can be evaluated by completing the square in the denominator.

$$\int \frac{dx}{a+bx+cx^2} = \frac{1}{c}\int \frac{dx}{\left(x+\dfrac{b}{2c}\right)^2 + \left(\dfrac{a}{c}-\dfrac{b^2}{4c^2}\right)}.$$

Positive-discriminant Case

Assume that the discriminant $q = b^2 - 4ac$ is positive. In that case, define u and A by,

$$u = x + \frac{b}{2c},$$

and

$$-A^2 = \frac{a}{c} - \frac{b^2}{4c^2} = \frac{1}{4c^2}\left(4ac - b^2\right).$$

The quadratic integral can now be written as,

$$\int \frac{dx}{a+bx+cx^2} = \frac{1}{c}\int \frac{du}{u^2 - A^2} = \frac{1}{c}\int \frac{du}{(u+A)(u-A)}.$$

The partial fraction decomposition,

$$\frac{1}{(u+A)(u-A)} = \frac{1}{2A}\left(\frac{1}{u-A} - \frac{1}{u+A}\right)$$

allows us to evaluate the integral:

$$\frac{1}{c}\int \frac{du}{(u+A)(u-A)} = \frac{1}{2Ac}\ln\left(\frac{u-A}{u+A}\right) + \text{constant}.$$

The final result for the original integral, under the assumption that $q > 0$, is

$$\int \frac{dx}{a+bx+cx^2} = \frac{1}{\sqrt{q}}\ln\left(\frac{2cx+b-\sqrt{q}}{2cx+b+\sqrt{q}}\right) + \text{constant, where } q = b^2 - 4ac.$$

Negative-discriminant Case

In case the discriminant $q = b^2 - 4ac$ is negative, the second term in the denominator in

$$\int \frac{dx}{a+bx+cx^2} = \frac{1}{c}\int \frac{dx}{\left(x+\dfrac{b}{2c}\right)^2 + \left(\dfrac{a}{c}-\dfrac{b^2}{4c^2}\right)}.$$

is positive. Then the integral becomes

$$\frac{1}{c}\int \frac{du}{u^2 + A^2}$$

$$= \frac{1}{cA}\int \frac{du / A}{(u / A)^2 + 1}$$

$$= \frac{1}{cA}\int \frac{dw}{w^2 + 1}$$

$$= \frac{1}{cA}\arctan(w) + \text{constant}$$

$$= \frac{1}{cA}\arctan\left(\frac{u}{A}\right) + \text{constant}$$

$$= \frac{1}{c\sqrt{\dfrac{a}{c} - \dfrac{b^2}{4c^2}}}\arctan\left(\frac{x + \dfrac{b}{2c}}{\sqrt{\dfrac{a}{c} - \dfrac{b^2}{4c^2}}}\right) + \text{constant}$$

$$= \frac{2}{\sqrt{4ac - b^2}}\arctan\left(\frac{2cx + b}{\sqrt{4ac - b^2}}\right) + \text{constant}.$$

Fresnel Integral

The Fresnel integrals $S(x)$ and $C(x)$ are two transcendental functions named after Augustin-Jean Fresnel that are used in optics and are closely related to the error function (erf). They arise in the description of near-field Fresnel diffraction phenomena and are defined through the following integral representations:

$$S(x) = \int_0^x \sin(t^2)\,dt, \quad C(x) = \int_0^x \cos(t^2)\,dt.$$

The simultaneous parametric plot of $S(x)$ and $C(x)$ is the Euler spiral (also known as the Cornu spiral or clothoid). Recently, they have been used in the design of highways and other engineering projects.

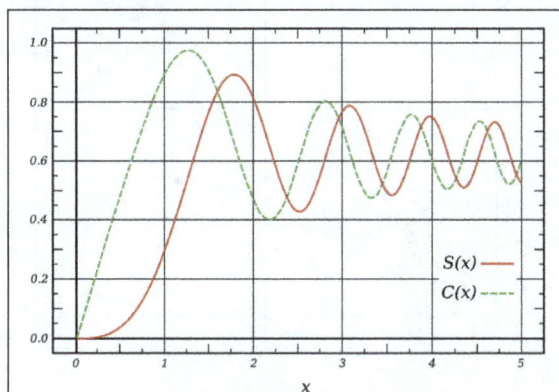

Plots of $S(x)$ and $C(x)$. The maximum of $C(x)$ is about 0.977451424. If the integrands of S and C were defined using $\pi t^2 / 2$ instead of t^2, then the image would be scaled vertically and horizontally.

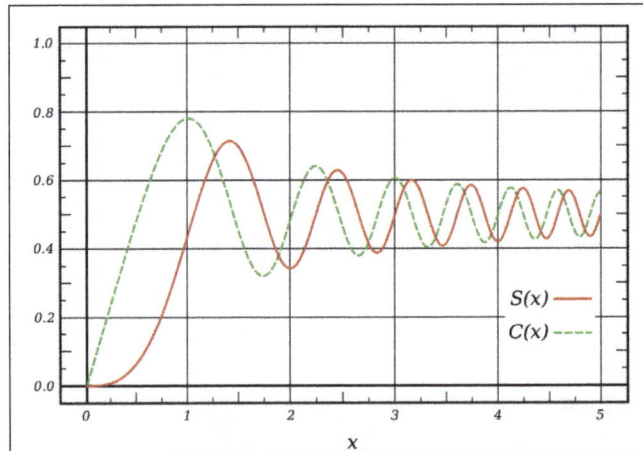

Fresnel integrals with arguments $\pi t^2/2$ instead of t^2 converge to 0.5.

The Fresnel integrals admit the following power series expansions that converge for all x:

$$S(x) = \int_0^x \sin(t^2)\,dt = \sum_{n=0}^{\infty} (-1)^n \frac{x^{4n+3}}{(2n+1)!(4n+3)}.$$

$$C(x) = \int_0^x \cos(t^2)\,dt = \sum_{n=0}^{\infty} (-1)^n \frac{x^{4n+1}}{(2n)!(4n+1)}.$$

Some authors, including Abramowitz and Stegun, by using above equations, $\pi t^2/2$ for the argument of the integrals defining $S(x)$ and $C(x)$. This changes their limits at infinity from $S(x)$ to $C(x)$ and the arc length for the first spiral turn (2π) to 2 (at $t=2$), all smaller by a factor $\sqrt{2/\pi}$. The alternative functions are also called Normalized Fresnel integrals.

Euler Spiral

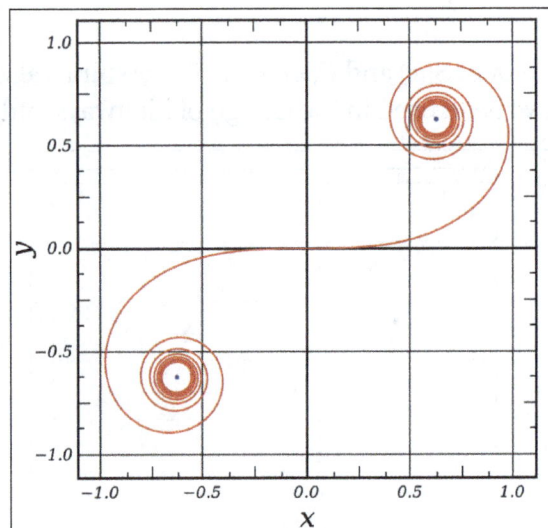

Euler spiral $(x, y) = (C(t), S(t))$. The spiral converges to the centre of the holes in the image as t tends to positive or negative infinity.

The Euler spiral, also known as Cornu spiral or clothoid, is the curve generated by a parametric plot of $S(t)$ against $C(t)$. The Cornu spiral was created by Marie Alfred Cornu as a nomogram for diffraction computations in science and engineering.

From the definitions of Fresnel integrals, the infinitesimals dx and dy are thus:

$$dx = C'(t)dt = \cos(t^2)dt,$$
$$dy = S'(t)dt = \sin(t^2)dt.$$

Thus the length of the spiral measured from the origin can be expressed as:

$$L = \int_0^{t_0} \sqrt{dx^2 + dy^2} = \int_0^{t_0} dt = t_0$$

That is, the parameter t is the curve length measured from the origin $(0,0)$, and the Euler spiral has infinite length. The vector $(\cos(t^2), \sin(t^2))$ also expresses the unit tangent vector along the spiral, giving $\theta = t^2$. Since t is the curve length, the curvature κ can be expressed as:

$$\kappa = \frac{1}{R} = \frac{d\theta}{dt} = 2t.$$

And the rate of change of curvature with respect to the curve length is,

$$\frac{d\kappa}{dt} = \frac{d^2\theta}{dt^2} = 2.$$

An Euler spiral has the property that its curvature at any point is proportional to the distance along the spiral, measured from the origin. This property makes it useful as a transition curve in highway and railway engineering: If a vehicle follows the spiral at unit speed, the parameter in the above derivatives also represents the time. That is, a vehicle following the spiral at constant speed will have a constant rate of angular acceleration.

Sections from Euler spirals are commonly incorporated into the shape of roller-coaster loops to make what are known as clothoid loops.

Properties

- $C(x)$ and $S(x)$ are odd functions of x.

- Asymptotics of the Fresnel integrals as $x \to \infty$ are given by the formulas:

$$S(x) = \sqrt{\frac{\pi}{2}}\left(\frac{\text{sign}(x)}{2} - \left[1 + O(x^{-4})\right]\left(\frac{\cos(x^2)}{x\sqrt{2\pi}} + \frac{\sin(x^2)}{x^3\sqrt{8\pi}}\right)\right),$$

$$C(x) = \sqrt{\frac{\pi}{2}}\left(\frac{\text{sign}(x)}{2} + \left[1 + O(x^{-4})\right]\left(\frac{\sin(x^2)}{x\sqrt{2\pi}} - \frac{\cos(x^2)}{x^3\sqrt{8\pi}}\right)\right).$$

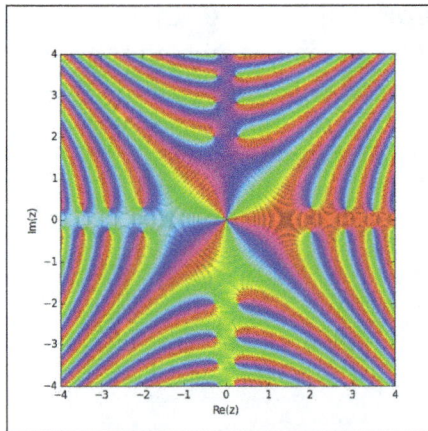

Complex Fresnel integral S(z)

- Using the power series expansions above, the Fresnel integrals can be extended to the domain of complex numbers, and they become analytic functions of a complex variable.

- The Fresnel integrals can be expressed using the error function as follows:

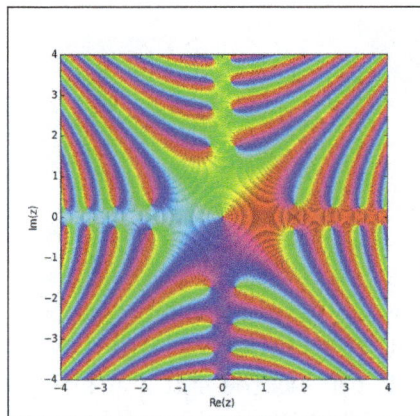

Complex Fresnel integral $C(z)$

$$S(z) = \sqrt{\frac{\pi}{2}} \frac{1+i}{4} \left[\operatorname{erf}\left(\frac{1+i}{\sqrt{2}} z \right) - i \operatorname{erf}\left(\frac{1-i}{\sqrt{2}} z \right) \right],$$

$$C(z) = \sqrt{\frac{\pi}{2}} \frac{1-i}{4} \left[\operatorname{erf}\left(\frac{1+i}{\sqrt{2}} z \right) + i \operatorname{erf}\left(\frac{1-i}{\sqrt{2}} z \right) \right].$$

or

$$C(z) + iS(z) = \sqrt{\frac{\pi}{2}} \frac{1+i}{2} \operatorname{erf}\left(\frac{1-i}{\sqrt{2}} z \right),$$

$$S(z) + iC(z) = \sqrt{\frac{\pi}{2}} \frac{1+i}{2} \operatorname{erf}\left(\frac{1+i}{\sqrt{2}} z \right).$$

- C and S are entire functions.

Limits as *x* Approaches Infinity

The integrals defining $C(x)$ and $S(x)$ cannot be evaluated in the closed form in terms of elementary functions, except in special cases. The limits of these functions as x goes to infinity are known:

$$\int_0^\infty \cos t^2 \, dt = \int_0^\infty \sin t^2 \, dt = \frac{\sqrt{2\pi}}{4} = \sqrt{\frac{\pi}{8}}.$$

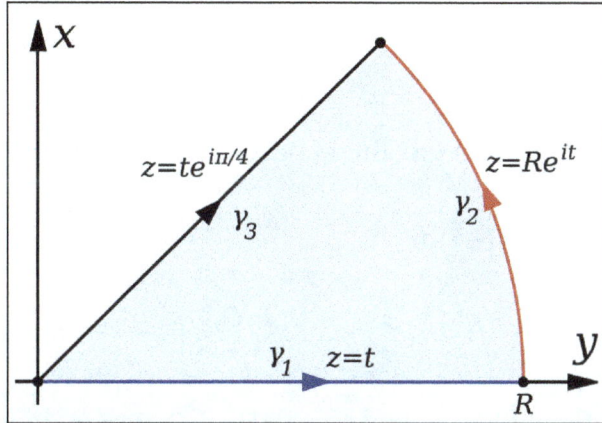

The sector contour used to calculate the limits of the Fresnel integrals

The limits of C and S as the argument tends to infinity can be found by the methods of complex analysis. This uses the contour integral of the function:

$$e^{-t^2}$$

around the boundary of the sector-shaped region in the complex plane formed by the positive x-axis, the bisector of the first quadrant $y = x$ with $x \geq 0$, and a circular arc of radius R centered at the origin.

As R goes to infinity, the integral along the circular arc γ_2 tends to 0,

$$\int_{\gamma_2} e^{-t^2} \, dt \leq \int_{\gamma_2} |e^{-t^2}| \, dt = R\int_0^{\pi/4} e^{-R^2 \cos 2t} \, dt \leq R\int_0^{\pi/4} e^{-R^2(1-\frac{4}{\pi}t)} \, dt = \frac{\pi}{4R}\left(1 - e^{-R^2}\right),$$

where polar coordinates $z = Re^{it}$ were used and Jordan's inequality was utilised for the second inequality. The integral along the real axis γ_1 tends to the half Gaussian integral:

$$\int_0^\infty e^{-t^2} \, dt = \frac{\sqrt{\pi}}{2}.$$

The integrand is an entire function on the complex plane, its integral along the whole contour is zero. Overall, we must have:

$$\int_0^\infty e^{-t^2} \, dt \int_{\gamma_3} e^{-t^2} \, dt$$

where γ_3 denotes the bisector of the first quadrant, as in the diagram. To evaluate the right hand side, parametrize the bisector as,

$$t = re^{\pi i/4} = \frac{\sqrt{2}}{2}(1+i)r$$

where r ranges from 0 to $+\infty$. The square of this expression is just $+ir^2$. Therefore, substitution gives the right hand side as:

$$\int_0^\infty e^{-ir^2}\frac{\sqrt{2}}{2}(1+i)dr.$$

Using Euler's formula to take real and imaginary parts of e^{-ir^2} gives this as:

$$\int_0^\infty (\cos(r^2) - i\sin(r^2))\frac{\sqrt{2}}{2}(1+i)dr$$

$$= \frac{\sqrt{2}}{2}\int_0^\infty \left[\cos(r^2) + \sin(r^2) + i\left(\cos(r^2) - \sin(r^2)\right)\right]dr = \frac{\sqrt{\pi}}{2} + 0i,$$

where we have written $0i$ to emphasize that the original Gaussian integral's value is completely real with zero imaginary part. Letting $I_C = \int_0^\infty \cos(r^2)dr, I_S = \int_0^\infty \sin(r^2)dr$ and then equating real and imaginary parts produces the following system of two equations in the two unknowns I_C, I_S:

$$I_C + I_S = \sqrt{\frac{\pi}{2}},$$

$$I_C - I_S = 0$$

Solving this for I_C and I_S gives the desired result.

Generalization

The integral,

$$\int x^m \exp(ix^n)dx = \int \sum_{l=0}^\infty \frac{i^l x^{m+nl}}{l!}dx = \sum_{l=0}^\infty \frac{i^l}{(m+nl+1)}\frac{x^{m+nl+1}}{l!}$$

is a confluent hypergeometric function and also an incomplete gamma function,

$$\int x^m \exp(ix^n)dx = \frac{x^{m+1}}{m+1}\,_1F_1\left(\begin{array}{c}\dfrac{m+1}{n}\\[2mm]1+\dfrac{m+1}{n}\end{array}\middle| ix^n\right)$$

$$= \frac{1}{n}i^{(m+1)/n}\gamma\left(\frac{m+1}{n}, -ix^n\right)$$

which reduces to Fresnel integrals if real or imaginary parts are taken:

$$\int x^m \sin(x^n)dx = \frac{x^{m+n+1}}{m+n+1}{}_1F_2\left(\begin{array}{c}\frac{1}{2}+\frac{m+1}{2n}\\[2mm]\frac{3}{2}+\frac{m+1}{2n},\frac{3}{2}\end{array}\middle|-\frac{x^{2n}}{4}\right).$$

The leading term in the asymptotic expansion is,

$${}_1F_1\left(\begin{array}{c}\frac{m+1}{n}\\[2mm]1+\frac{m+1}{n}\end{array}\middle|ix^n\right) \sim \frac{m+1}{n}\Gamma\left(\frac{m+1}{n}\right)e^{i\pi(m+1)/(2n)}x^{-m-1},$$

and therefore

$$\int_0^\infty x^m \exp(ix^n)dx = \frac{1}{n}\Gamma\left(\frac{m+1}{n}\right)e^{i\pi(m+1)/(2n)}.$$

For $m = 0$, the imaginary part of this equation in particular is,

$$\int_0^\infty \sin(x^a)dx = \Gamma\left(1+\frac{1}{a}\right)\sin\left(\frac{\pi}{2a}\right),$$

with the left-hand side converging for $a > 1$ and the right-hand side being its analytical extension to the whole plane less where lie the poles of $\Gamma(a^{-1})$.

The Kummer transformation of the confluent hypergeometric function is,

$$\int x^m \exp(ix^n)dx = V_{n,m}(x)e^{ix^n},$$

with

$$V_{n,m} := \frac{x^{m+1}}{m+1}{}_1F_1\left(\begin{array}{c}1\\[2mm]1+\frac{m+1}{n}\end{array}\middle|-ix^n\right).$$

Numerical Approximation

For computation to arbitrary precision, the power series is suitable for small argument. For large argument, asymptotic expansions converge faster. Continued fraction methods may also be used.

For computation to particular target precision, other approximations have been developed. Cody developed a set of efficient approximations based on rational functions that give relative errors down to 2×10^{-19}. A fortran implementation of the Cody approximation that includes the values of the coefficients needed for implementation in other languages was published by van Snyder. Boersma developed an approximation with error less than 1.6×10^{-9}.

Borwein Integral

In mathematics, a Borwein integral is an integral involving products of sinc(ax), where the sinc function is given by $\sin c(x) = \dfrac{\sin(x)}{x}$ for x not equal to 0, and sinc(0) = 1.

These integrals are remarkable for exhibiting apparent patterns which, however, eventually break down. An example is as follows,

$$\int_0^\infty \frac{\sin(x)}{x} dx = \pi/2$$

$$\int_0^\infty \frac{\sin(x)}{x} \frac{\sin(x/3)}{x/3} dx = \pi/2$$

$$\int_0^\infty \frac{\sin(x)}{x} \frac{\sin(x/3)}{x/3} \frac{\sin(x/5)}{x/5} dx = \pi/2$$

This pattern continues up to

$$\int_0^\infty \frac{\sin(x)}{x} \frac{\sin(x/3)}{x/3} \cdots \frac{\sin(x/13)}{x/13} dx = \pi/2$$

Nevertheless, at the next step the obvious pattern fails,

$$\int_0^\infty \frac{\sin(x)}{x} \frac{\sin(x/3)}{x/3} \cdots \frac{\sin(x/15)}{x/15} dx = \frac{467807924713440738696537864469}{935615849440640907310521750000} \pi$$

$$= \frac{\pi}{2} - \frac{6879714958723010531}{935615849440640907310521750000} \pi$$

$$\simeq \frac{\pi}{2} - 2.31 \times 10^{-11}$$

In general, similar integrals have value $\dfrac{\pi}{2}$ whenever the numbers 3, 5, 7... are replaced by positive real numbers such that the sum of their reciprocals is less than 1.

In the example above, $\dfrac{1}{3} + \dfrac{1}{5} + \ldots + \dfrac{1}{13} < 1$, but $\dfrac{1}{3} + \dfrac{1}{5} + \ldots + \dfrac{1}{5} > 1$

An example for a longer series,

$$\int_0^\infty 2\cos(x) \frac{\sin(x)}{x} \frac{\sin(x/3)}{x/3} \cdots \frac{\sin(x/111)}{x/111} dx = \pi/2,$$

$$\int_0^\infty 2\cos(x) \frac{\sin(x)}{x} \frac{\sin(x/3)}{x/3} \cdots \frac{\sin(x/111)}{x/111} \frac{\sin(x/113)}{x/113} dx < \pi/2$$

is shown in together with an intuitive mathematical explanation of the reason why the original and the extended series break down.

In this case, $\dfrac{1}{3} + \dfrac{1}{5} + \ldots + \dfrac{1}{111} < 2$ but $\dfrac{1}{3} + \dfrac{1}{5} + \ldots + \dfrac{1}{113} > 2.$

General Formula

Given a sequence of real numbers, $a_0, a_1, a_2, \ldots,$, a general formula for the integral

$$\int_0^\infty \prod_{k=0}^n \frac{\sin(a_k x)}{a_k x} \, dx$$

can be given. To state the formula, we will need to consider sums involving the a_k. In particular, if $\gamma = (\gamma_1, \gamma_2, \ldots, \gamma_n) \in \{\pm 1\}^n$. is an n-tuple where each entry is ± 1, then we write $b_\gamma = a_0 + \gamma_1 a_1 + \gamma_2 a_2 + \cdots + \gamma_n a_n$, which is a kind of alternating sum of the first few a_k, and we set:

$$\int_0^\infty \prod_{k=0}^n \frac{\sin(a_k x)}{a_k x} dx = \frac{\pi}{2 a_0} C_n$$

where,

$$C_n = \frac{1}{2^n n! \prod_{k=1}^n a_k} \sum_{\gamma \in \{\pm 1\}^n} \epsilon_\gamma b_\gamma^n \, \mathrm{sgn}(b_\gamma)$$

In the case when $a_0 > |a_1| + |a_2| + \cdots + |a_n|$, we have $C_n = 1.$

Furthermore, if there is an n so that for each $k = 0, \ldots, n-1$ we have $0 < a_n < 2 a_k$ and $a_1 + a_2 + \cdots + a_{n-1} < a_0 < a_1 + a_2 + \cdots + a_{n-1} + a_n$, which means that n is the first value when the partial sum of the first n elements of the sequence exceed a_0 then $C_k = 1$ for each $k = 0, \ldots, n-1$ but,

$$C_n = 1 - \frac{(a_1 + a_2 + \cdots + a_n - a_0)^n}{2^{n-1} n! \prod_{k=1}^n a_k}$$

The first example is the case when $a_k = \dfrac{1}{2k+1}$. Note that if $n = 7$ then $a_7 = \dfrac{1}{15}$ and $\dfrac{1}{3} + \dfrac{1}{5} + \dfrac{1}{7} + \dfrac{1}{9} + \dfrac{1}{11} + \dfrac{1}{13} \approx 0.955$ but $\dfrac{1}{3} + \dfrac{1}{5} + \dfrac{1}{7} + \dfrac{1}{9} + \dfrac{1}{11} + \dfrac{1}{13} + \dfrac{1}{15} \approx 1.02$, so because $a_0 = 1$, we get that,

$$\int_0^\infty \frac{\sin(x)}{x} \frac{\sin(x/3)}{x/3} \cdots \frac{\sin(x/13)}{x/13} dx = \frac{\pi}{2}$$

which remains true if we remove any of the products but that,

$$\int_0^\infty \frac{\sin(x)}{x}\frac{\sin(x/3)}{x/3}\ldots\frac{\sin(x/15)}{x/15}dx = \frac{\pi}{2}\left(1-\frac{(3^{-1}+5^{-1}+7^{-1}+9^{-1}+11^{-1}+13^{-1}+15^{-1}-1)^7}{2^6\cdot 7!\cdot(1/3\cdot1/5\cdot1/7\cdot1/9\cdot1/11\cdot1/13\cdot1/15)}\right)$$

which is equal to the value given previously.

Dirichlet Integral

There are several types of integrals which go under the name of a "Dirichlet integral." The integral

$$D[u] = \int_\Omega |\nabla u|^2 dV$$

appears in Dirichlet's principle.

The integral,

$$\frac{1}{2\pi}\int_{-\pi}^{\pi} f(x)\frac{\sin\left[\left(n+\frac{1}{2}\right)x\right]}{\sin\left(\frac{1}{2}x\right)}dx,$$

where the kernel is the Dirichlet kernel, gives the nth partial sum of the Fourier series.

Another integral is denoted,

$$\delta_k \equiv \frac{1}{\pi}\int_{-\infty}^{\infty}\frac{\sin\alpha_k\rho_k}{\rho_k}e^{i\rho_k\gamma_k}\,d\,\rho_k \begin{cases}0 & \text{for } |\gamma_k| > \alpha_k \\ 1 & \text{for } |\gamma_k| < \alpha_k\end{cases}$$

for $k = 1, \ldots, n$.

There are two types of Dirichlet integrals which are denoted using the letters and J. The type 1 Dirichlet integrals are denoted I, J, and IJ, and the type 2 Dirichlet integrals are denoted C, D, and CD.

The type 1 integrals are given by,

$$I \equiv \iint\ldots\int f(t_1 + t_2 + \ldots + t_n)\,t_1^{\alpha_1-1}\,t_2^{\alpha_2-1}\ldots t_n^{\alpha_n-1}\,d\,t_2\ldots d\,t_n$$
$$= \frac{\Gamma(\alpha_1)\Gamma(\alpha_2)\ldots\Gamma(\alpha_n)}{\Gamma(\Sigma_n\alpha_n)}\int_0^1 f(\tau)\tau^{(\Sigma_n\alpha)-1}\,d\tau$$

where $\Gamma(z)$ is the gamma function. In the case $n = 2$,

$$I = \iint_T x^p y^q\,dx\,dy = \frac{p!q!}{(p+q+2)} = \frac{B(p+1, q+1)}{p+q+2}$$

where the integration is over the triangle T bounded by the x-axis, y-axis, and line $x + y = 1$ and B(x,y) is the beta function.

The type 2 integrals are given for b -D vectors a and r, and $0 \le c \le b$,

$$C_a^{(b)}(r, m) = \frac{\Gamma(m + R)}{\Gamma(m) \prod_{i=1}^{\phi} \Gamma(r_i)} \int_0^{a_1} \cdots \int_0^{a_b} \frac{\prod_{i=1}^{\phi} x_i^{r_i - 1} dx_i}{\left(1 + \sum_{i=1}^{b} x_i\right)^{m + R}}$$

$$D_a^{(b)}(r, m) = \frac{\Gamma(m + R)}{\Gamma(m) \prod_{i=1}^{\phi} \Gamma(r_i)} \int_{a_1}^{\infty} \cdots \int_{a_k}^{-\infty} \frac{\prod_{i=1}^{\phi} x_i^{r_i - 1} dx_i}{\left(1 + \sum_{i=1}^{b} x_i\right)^{m + R}}$$

$$CD_a^{(c, d-c)}(r, m) = \frac{\Gamma(m + R)}{\Gamma(m) \prod_{i=1}^{\phi} \Gamma(r_i)} \int_0^{a_c} \int_{a_{c+1}}^{\infty} \cdots \int_{a_b}^{\infty} \frac{\prod_{i=1}^{\phi} x_i^{r_i - 1} dx_i}{\left(1 + \sum_{i=1}^{b} x_i\right)^{m + R}}$$

Where,

$$R \equiv \sum_{i=1}^{k} r_i$$

$$a_i \equiv \frac{p_i}{1 - \sum_{i=1}^{k} p_i},$$

and p_i are the cell probabilities. For equal probabilities, $a_i = 1$. The Dirichlet D integral can be expanded as a multinomial series as,

$$D_a^{(b)}(r, m) = \frac{1}{\left(1 + \sum_{i=1}^{b}\right)^m} \sum_{x_1 < r_1} \cdots \sum_{x_b < r_b} \binom{m - 1 \sum_{a=1}^{b} x_i}{m - 1, x_1, \ldots, x_b}$$

$$\prod_{i=1}^{b} \left(\frac{a_i}{1 + \sum_{k=1}^{b} a_k}\right)^{x_i}.$$

For small b, C and D can be expressed analytically either partially or fully for general arguments and a$_i$=1.

$$C_1^{(1)}(r_2; r_1) = \frac{\Gamma(r_1 + r_2) {}_2F_1(r_2, r_1 + r_2; 1 + r_2; -1)}{r_2 \Gamma(r_1) \Gamma(r_2)}$$

$$C_1^{(2)}(r_2, r_3\,;\,r_1) = \frac{\Gamma(r_1 + r_2 + r_3)}{r_2\Gamma(r_1)\Gamma(r_2)\,\Gamma(r_3)} \int_0^1 {}_2F_1\, y^{r_3-1}\,(1+y)^{-(r_1+r_2+r_3)}\,dy,$$

Where,

$$2F_1 \equiv {}_2F_1\left(r_2, r_1 + r_2 + r_3\,;\,1 + r_2, -(1+y)^{-1}\right)$$

is a hypergeometric function.

$$D_1^{(2)}(r_2, r_3\,;\,r_1) = \frac{\Gamma(r_1 + r_2 + r_3)}{(r_1 + r_3)\,\Gamma(r_1)\Gamma(r_2)\Gamma(r_3)} \int_1^\infty {}_2F_1\, y^{r_3-1}\,dy,$$

Where,

$${}_2F_1 = {}_2F_1(r_1 + r_3, r_1 + r_2 + r_3\,;\,1 + r_1 + r_3; -1-y)\,.$$

References

- Complex Variables, M.R. Speigel, S. Lipschutz, J.J. Schiller, D. Spellman, Schaum's Outlines Series, McGraw Hill (USA), 2009, ISBN 978-0-07-161569-3

- Differential-calculus: encyclopediaofmath.org, Retrieved 25 April, 2019

- Wheeler, Gerard F.; Crummett, William P. (1987). "The Vibrating String Controversy". Am. J. Phys. 55 (1): 33–37. Bibcode:1987AmJPh..55...33W. doi:10.1119/1.15311

- Integral-calculus: encyclopediaofmath.org, Retrieved 14 June, 2019

- Advanced Calculus (3rd edition), R. Wrede, M.R. Spiegel, Schaum's Outline Series, 2010, ISBN 978-0-07-162366-7

- DirichletIntegrals: mathworld.wolfram.com, Retrieved 27 July, 2019

Multivariable Calculus

The branch of calculus which deals with the differentiation and integration of functions involving several variables instead of one is known as multivariable calculus. A definite integral which is a function of more than one real variable is known as a multiple integral. This chapter discusses in detail these theories and methodologies related to multivariable calculus.

Multivariable calculus is the study of calculus with more than one variable. We understand differentiation and integration of two or more variable by partial derivative by using the first order of test in finding the critical point. Then we apply the second order of test to find maxima, minima and saddle point. We solve optimization problems using the first and second order.

Multivariable calculus is a branch of calculus in one variable to calculus with functions of more than one variable. In single variable calculus, we study the function of single variable whereas in multivariable calculus we study with two or more variables.

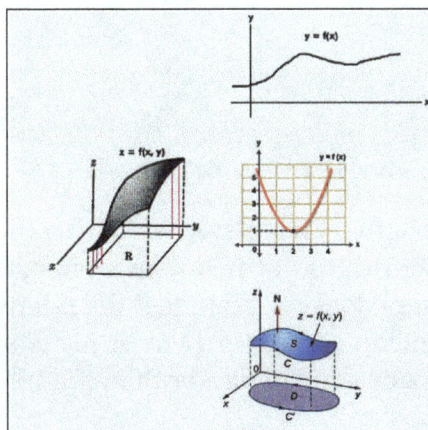

Partial Derivatives: It is derivative of a function of two or more variables with respect to one of those variables, with the other held constant. It is used in vector calculus and differential geometry.

Let's assume F(x,y) to be a function with two variables. By keeping y constant and differentiable F (assuming F is differentiable) with respect to variable x. What we obtain is the partial derivative of F with respect to x and is denoted by $\partial F/\partial x$ OR Fx.

In the same way, partial derivative of F with respect to y is denoted by $\partial F/\partial y$ OR Fy.

Critical point of function of two variables: Critical point of a function with two variables is a point where the partial derivative of first order are equal to 0. To find a critical point we must first take the derivative of the function. Then set that derivative equal to 0 and solve for x. Each value of x that w get is known as the critical number.

Let's find the critical point of function F defined by F(x,y)= x² + y².

We start with finding the first order partial derivative.

F_x(x,y)= 2x

F_y(x,y)= 2y

Now we will solve the equation F_x(x,y)=0 and F_y(x,y)= 0 simultaneously.

F_x(x,y) = 2x = 0

F_y(x,y) = 2y = 0

The solution of the above equation is the ordered pair (0,0).

The graph of F(x,y)= x² + y² states that at the critical point (0,0) f has a minimum value.

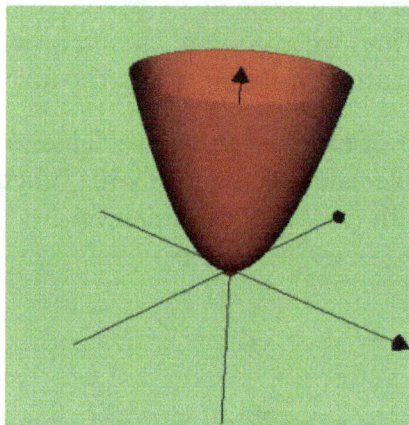

Maxima and minima of functions of two variable: After getting the critical point we do the second derivative test to determine if a critical point is a relative maximum or relative minimum. The meaning of maximum and minimum does not state that the relative maximum or minimum is the largest/smallest value that the function will ever take. It just states that in some region around point (a,b) the function will always be smaller/larger than F(a,b). It is possible for the function to be larger/smaller outside of that region.

Firstly, we need to figure out how many second derivatives we have,

$$\frac{\partial^2 F}{\partial x^2} = F_{xx}$$

$$F_{xy} = \frac{\partial^2 F}{\partial_x \partial y}$$

$$\frac{\partial^2 F}{\partial y \partial x} = F_{yx}$$

$$\frac{\partial^2 F}{\partial y^2} = F_{yy}$$

It is interesting to note that,

$$F_{xy} = \frac{\partial^2 F}{\partial_x \partial y} = \frac{\partial^2 F}{\partial y \partial x} = F_{yx}$$

The second derivative test states that when we have a critical point (x_0, y_0) of Function of two variables and have to calculate the partial derivative.

Let $A = F_{xx}(x_0, y_0), B = F_{xy}(x_0, y_0), C = F_{yy}(x_0, y_0)$

If $AC - B^2 > 0$ and $A > 0$ then it is the minimum and when $A < 0$ is the maximum.

When, $AC - B^2 < 0$ then it is called the saddle point.

And when, $AC - B^2 = 0$ then we cannot conclude whether it will be the minimum, maximum or the saddle point.

Optimization problems with functions of two variables: Optimization problems involve optimizing functions in two variables using first and second order.

Let us look at some problems closely.

Example: $F(x,y) = x^2 + 2y^2 - x^2y$

Solution: Critical point occurs where F_x and F_y are simultaneously 0.

$F_x = 2x - 2xy = 2x(1-y)$

$F_y = 4y - x^2$

$F_x = 0$ if $x = 0$ or $y = 1$

Using this in the equation $F_y = 0$

If $x = 0, y = 0$

If $y = 1$ then $4 - x^2 = 0$

So, $x =$

Therefore, we have $(0,0)$, $(2,1)$, and $(-2,1)$

Now using the second partial test to classify

$D = F_{xx} \cdot F_{yy} - (F_{xy})^2 = (2 - 2y).(4) - (-2x)^2$

At $(0,0)$ D= 8 and $F_{xx} = 2$ therefore we have a minimum.

At $(2,1)$ D = -16 is the saddle point.

At $(-2,1)$ D = -16 = 0 therefore is a saddle point.

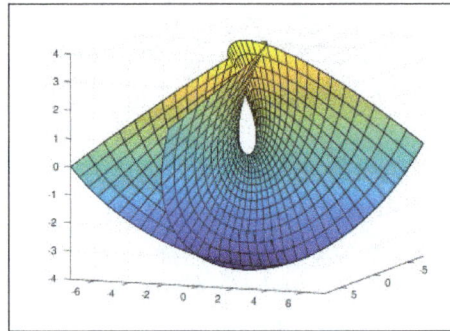

Application of Multivariable Calculus: Multivariable calculus is useful considering that most natural phenomenon are non-linear and can be best described by using multivariable calculus and differential equation. For example relationship between speed, position and acceleration can be defined by multivariable calculus and differential equation.

MULTIPLE INTEGRAL

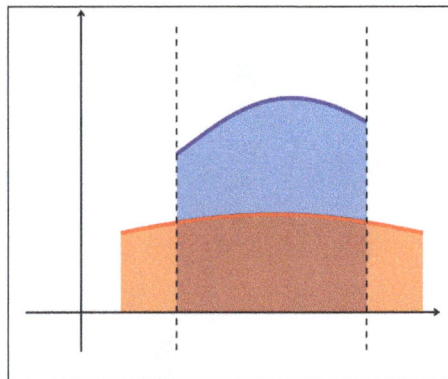

Integral as area between two curves.

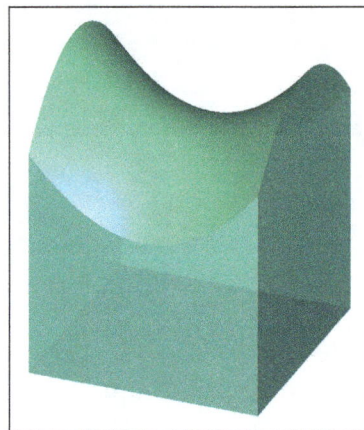

Double integral as volume under a surface $z = 10 - \dfrac{x^2 - y^2}{8}$. The rectangular region at the bottom of the body is the domain of integration, while the surface is the graph of the two-variable function to be integrated.

The multiple integral is a definite integral of a function of more than one real variable, for example, $f(x, y)$ or $f(x, y, z)$. Integrals of a function of two variables over a region in R² are called double integrals, and integrals of a function of three variables over a region of R³ are called triple integrals.

Just as the definite integral of a positive function of one variable represents the area of the region between the graph of the function and the x-axis, the double integral of a positive function of two variables represents the volume of the region between the surface defined by the function (on the three-dimensional Cartesian plane where $z = f(x, y)$ and the plane which contains its domain. If there are more variables, a multiple integral will yield hypervolumes of multidimensional functions.

Multiple integration of a function in n variables: $f(x_1, x_2, ..., x_n)$ over a domain D is most commonly represented by nested integral signs in the reverse order of execution (the leftmost integral sign is computed last), followed by the function and integrand arguments in proper order (the integral with respect to the rightmost argument is computed last). The domain of integration is either represented symbolically for every argument over each integral sign, or is abbreviated by a variable at the rightmost integral sign:

$$\int \cdots \int_D f(x_1, x_2, \ldots, x_n) dx_1 \cdots dx_n$$

Since the concept of an antiderivative is only defined for functions of a single real variable, the usual definition of the indefinite integral does not immediately extend to the multiple integral.

Mathematical Definition

For $n > 1$, consider a so-called "half-open" n-dimensional hyperrectangular domain T, defined as:

$$T = [a_1, b_1) \times [a_2, b_2) \times \cdots \times [a_n, b_n) \subseteq R^n.$$

Partition each interval $[a_j, b_j)$ into a finite family I_j of non-overlapping subintervals i_{ja}, with each subinterval closed at the left end, and open at the right end.

Then the finite family of subrectangles C given by

$$C = I_1 \times I_2 \times \cdots \times I_n$$

is a partition of T; that is, the subrectangles C_k are non-overlapping and their union is T.

Let $f : T \to R$ be a function defined on T. Consider a partition C of T as defined above, such that C is a family of m subrectangles C_m and

$$T = C_1 \cup C_2 \cup \cdots \cup C_m$$

We can approximate the total $(n + 1)$th-dimensional volume bounded below by the n-dimensional hyperrectangle T and above by the n-dimensional graph of f with the following Riemann sum:

$$\sum_{k=1}^{m} f(P_k) m(C_k)$$

where P_k is a point in C_k and $m(C_k)$ is the product of the lengths of the intervals whose Cartesian product is C_k, also known as the measure of C_k.

The diameter of a subrectangle C_k is the largest of the lengths of the intervals whose Cartesian product is C_k. The diameter of a given partition of T is defined as the largest of the diameters of the subrectangles in the partition. Intuitively, as the diameter of the partition C is restricted smaller and smaller, the number of subrectangles m gets larger, and the measure $m(C_k)$ of each subrectangle grows smaller. The function f is said to be Riemann integrable if the limit:

$$S = \lim_{\delta \to 0} \sum_{k=1}^{m} f(P_k) m(C_k)$$

exists, where the limit is taken over all possible partitions of T of diameter at most δ.

If f is Riemann integrable, S is called the Riemann integral of f over T and is denoted

$$\int \cdots \int_T f(x_1, x_2, \ldots, x_n) dx_1 \cdots dx_n$$

Frequently this notation is abbreviated as

$$\int_T f(\mathrm{x}) d^n \mathrm{x}.$$

where x represents the n-tuple $(x_1, \ldots x_n)$ and d^nx is the n-dimensional volume differential.

The Riemann integral of a function defined over an arbitrary bounded n-dimensional set can be defined by extending that function to a function defined over a half-open rectangle whose values are zero outside the domain of the original function. Then the integral of the original function over the original domain is defined to be the integral of the extended function over its rectangular domain, if it exists.

In what follows the Riemann integral in n dimensions will be called the multiple integral.

Properties

Multiple integrals have many properties common to those of integrals of functions of one variable (linearity, commutativity, monotonicity, and so on). One important property of multiple integrals is that the value of an integral is independent of the order of integrands under certain conditions. This property is popularly known as Fubini's theorem.

Particular Cases

In the case of $T \subseteq R^2$, the integral

$$l = \iint_T f(x, y) \, dx \, dy$$

is the double integral of f on T, and if $T \subseteq R^3$ the integral,

$$l = \iiint_T f(x, y, z) \, dx \, dy \, dz$$

is the triple integral of f on T.

Notice that, by convention, the double integral has two integral signs, and the triple integral has three; this is a notational convention which is convenient when computing a multiple integral as an iterated integral.

Methods of Integration

The resolution of problems with multiple integrals consists, in most cases, of finding a way to reduce the multiple integral to an iterated integral, a series of integrals of one variable, each being directly solvable. For continuous functions, this is justified by Fubini's theorem. Sometimes, it is possible to obtain the result of the integration by direct examination without any calculations.

The following are some simple methods of integration:

Integrating Constant Functions

When the integrand is a constant function c, the integral is equal to the product of c and the measure of the domain of integration. If $c = 1$ and the domain is a subregion of R^2, the integral gives the area of the region, while if the domain is a subregion of R^3, the integral gives the volume of the region.

Example. Let $f(x, y) = 2$ and

$$D = \left\{(x, y) \in R^2 : 2 \le x \le 4 \,; 3 \le y \le 6\right\}$$

in which case

$$\int_3^6 \int_2^4 2 \, dx\, dy = 2 \int_3^6 \int_2^4 1 \, dx\, dy = 2 \cdot \text{area}(D) = 2 \cdot (2 \cdot 3) = 12,$$

since by definition we have:

$$\int_3^6 \int_2^4 1 \, dx\, dy = \text{area}(D).$$

Use of Symmetry

When the domain of integration is symmetric about the origin with respect to at least one of the variables of integration and the integrand is odd with respect to this variable, the integral is equal to zero, as the integrals over the two halves of the domain have the same absolute value but opposite signs. When the integrand is even with respect to this variable, the integral is equal to twice the integral over one half of the domain, as the integrals over the two halves of the domain are equal.

Example: Consider the function $f(x, y) = 2 \sin(x) - 3y^3 + 5$ integrated over the domain

$$T = \left\{(x, y) \in R^2 : x^2 + y^2 \le 1\right\},$$

a disc with radius 1 centered at the origin with the boundary included.

Using the linearity property, the integral can be decomposed into three pieces:

$$\iint_T \left(2\sin x - 3y^3 + 5\right)\, dx\, dy = \iint_T 2\sin x\, dx\, dy - \iint_T 3y^3\, dx\, dy + \iint_T 5\, dx\, dy$$

The function $2\sin(x)$ is an odd function in the variable x and the disc T is symmetric with respect to the y-axis, so the value of the first integral is 0. Similarly, the function $3y^3$ is an odd function of y, and T is symmetric with respect to the x-axis, and so the only contribution to the final result is that of the third integral. Therefore the original integral is equal to the area of the disk times 5, or 5π.

Example: Consider the function $f(x, y, z) = x\, exp\left(y^2 + z^2\right)$ and as integration region the sphere with radius 2 centered at the origin,

$$T = \left\{(x, y, z) \in \mathbb{R}^3 : x^2 + y^2 + z^2 \leq 4\right\}.$$

The "ball" is symmetric about all three axes, but it is sufficient to integrate with respect to x-axis to show that the integral is 0, because the function is an odd function of that variable.

Normal Domains on R²

This method is applicable to any domain D for which:

- The projection of D onto either the x-axis or the y-axis is bounded by the two values, a and b.

- Any line perpendicular to this axis that passes between these two values intersects the domain in an interval whose endpoints are given by the graphs of two functions, α and β.

Such a domain will be here called a *normal domain*. Elsewhere in the literature, normal domains are sometimes called type I or type II domains, depending on which axis the domain is fibred over. In all cases, the function to be integrated must be Riemann integrable on the domain, which is true (for instance) if the function is continuous.

x-axis

If the domain D is normal with respect to the x-axis, and $f : D \to \mathbb{R}$ is a continuous function; then $\alpha(x)$ and $\beta(x)$ (both of which are defined on the interval $[a, b]$) are the two functions that determine D. Then, by Fubini's theorem:

$$\iint_D f(x, y)\, dx\, dy = \int_a^b dx \int_{\alpha(x)}^{\beta(x)} f(x, y)\, dy.$$

y-axis

If D is normal with respect to the y-axis and $f : D \to \mathbb{R}$ is a continuous function; then $\alpha(y)$ and $\beta(y)$ (both of which are defined on the interval $[a, b]$) are the two functions that determine D. Again, by Fubini's theorem:

$$\iint_D f(x, y)\, dx\, dy \quad \int_a^b dy \int_{(y)}^{(y)} f(x, y)\, dx.$$

Normal Domains on R³

If T is a domain that is normal with respect to the xy-plane and determined by the functions $\alpha(x, y)$ and $\beta(x, y)$, then

$$\iiint_T f(x,y,z) \, dx \, dy \, dz = \iint_D \int_{\alpha(x,y)}^{\beta(x,y)} f(x,y,z) \, dz \, dx \, dy$$

This definition is the same for the other five normality cases on R³. It can be generalized in a straightforward way to domains in R^n.

Change of Variables

The limits of integration are often not easily interchangeable (without normality or with complex formulae to integrate). One makes a change of variables to rewrite the integral in a more "comfortable" region, which can be described in simpler formulae. To do so, the function must be adapted to the new coordinates.

Example: The function is $f(x,y) = (x-1)^2 + \sqrt{y}$; if one adopts the substitution $x' = x - 1$, $y' = y$ therefore $x = x' + 1$, $y = y'$ one obtains the new function $f_2(x,y) = (x')^2 + \sqrt{y}$.

- Similarly for the domain because it is delimited by the original variables that were transformed before (x and y in example).

- the differentials dx and dy transform via the absolute value of the determinant of the Jacobian matrix containing the partial derivatives of the transformations regarding the new variable.

There exist three main "kinds" of changes of variable (one in R², two in R³); however, more general substitutions can be made using the same principle.

Polar Coordinates

In R² if the domain has a circular symmetry and the function has some particular characteristics one can apply the *transformation to polar coordinates* which means that the generic points $P(x, y)$ in Cartesian coordinates switch to their respective points in polar coordinates. That allows one to change the shape of the domain and simplify the operations.

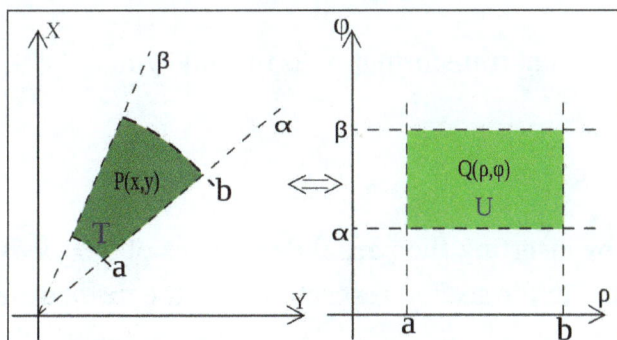

Transformation from cartesian to polar coordinates.

The fundamental relation to make the transformation is the following:

$$f(x, y) \rightarrow f(\rho \cos \varphi, \rho \sin \varphi).$$

Example: The function is $f(x, y) = x + y$ and applying the transformation one obtains,

$$f(\rho, \varphi) = \rho \cos \varphi + \rho \sin \varphi = \rho(\cos \varphi + \sin \varphi).$$

Example: The function is $f(x, y) = x^2 + y^2$, in this case one has:

$$f(\rho, \varphi) = \rho^2 \left(\cos^2 \varphi + \sin^2 \varphi \right) = \rho^2$$

using the Pythagorean trigonometric identity (very useful to simplify this operation).

The transformation of the domain is made by defining the radius' crown length and the amplitude of the described angle to define the ρ, φ intervals starting from x, y.

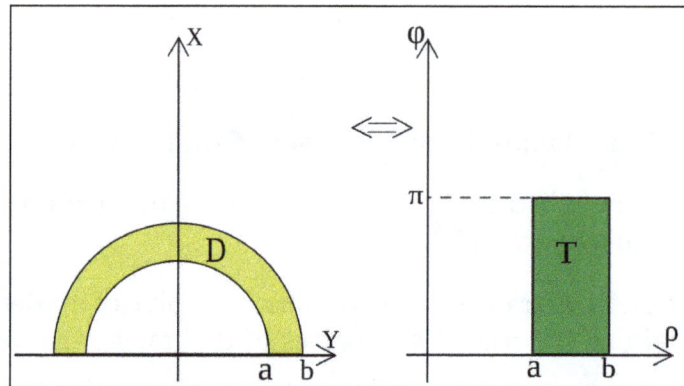

Example of a domain transformation from cartesian to polar.

Example: The domain is $D = \{x^2 + y^2 \leq 4\}$, that is a circumference of radius 2; it's evident that the covered angle is the circle angle, so φ varies from 0 to 2π, while the crown radius varies from 0 to 2 (the crown with the inside radius null is just a circle).

Example: The domain is $D = \{x^2 + y^2 \leq 9, x^2 + y^2 \geq 4, y \geq 0\}$, that is the circular crown in the positive y half-plane; φ describes a plane angle while ρ varies from 2 to 3. Therefore the transformed domain will be the following rectangle:

$$T = \{2 \leq \rho \leq 3, 0 \leq \varphi \leq \pi\}.$$

The Jacobian determinant of that transformation is the following:

$$\frac{\partial(x, y)}{\partial(\rho, \varphi)} = \begin{vmatrix} \cos \varphi & -\rho \sin \varphi \\ \sin \varphi & \rho \cos \varphi \end{vmatrix} = \rho$$

which has been obtained by inserting the partial derivatives of $x = \rho cos(\varphi), y = \rho sin(\varphi)$ in the first column respect to ρ and in the second respect to φ, so the $dx\, dy$ differentials in this transformation become $\rho\, d\rho\, d\varphi$.

Once the function is transformed and the domain evaluated, it is possible to define the formula for the change of variables in polar coordinates:

$$\iint_D f(x,y)dx\,dy = \iint_T f(\rho\cos\varphi, \rho\sin\varphi)\rho\,d\rho\,d\varphi.$$

φ is valid in the [0, 2π] interval while ρ, which is a measure of a length, can only have positive values.

Example: The function is $f(x, y) = x$ and the domain is the same as in previous example. From the previous analysis of D we know the intervals of ρ (from 2 to 3) and of φ (from 0 to π). Now we change the function:

$$f(x,y) = x \rightarrow f(\rho,\varphi) = \rho\cos\varphi.$$

finally let's apply the integration formula:

$$\iint_D x\,dx\,dy = \iint_T \rho\cos\varphi\rho\,d\rho\,d\varphi.$$

Once the intervals are known, you have

$$\int_0^\pi \int_2^3 \rho^2\cos\varphi\,d\rho\,d\varphi = \int_0^\pi \cos\varphi\,d\varphi\left[\frac{\rho^3}{3}\right]_2^3 = [\sin\varphi]_0^\pi\left(9 - \frac{8}{3}\right) = 0.$$

Cylindrical Coordinates

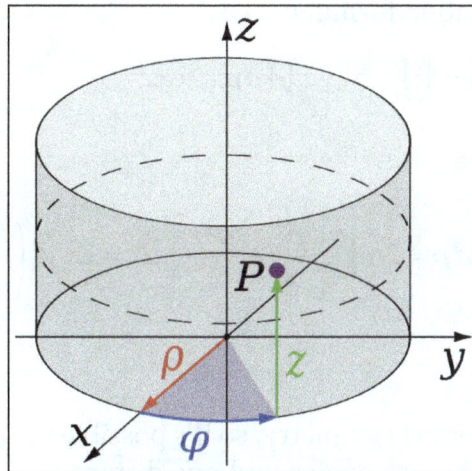

Cylindrical coordinates.

In R³ the integration on domains with a circular base can be made by the *passage to cylindrical coordinates*; the transformation of the function is made by the following relation:

$$f(x,y,z) \rightarrow f(\rho\cos\varphi, \rho\sin\varphi, z)$$

The domain transformation can be graphically attained, because only the shape of the base varies, while the height follows the shape of the starting region.

Example: The region is $D = \{x^2 + y^2 \leq 9, x^2 + y^2 \geq 4, 0 \leq z \leq 5\}$ if the transformation is applied, this region is obtained:

$$T = \{2 \leq \rho \leq 3, 0 \leq \varphi \leq 2\pi, 0 \leq z \leq 5\}$$

Because the z component is unvaried during the transformation, the dx dy dz differentials vary as in the passage to polar coordinates: therefore, they become $\rho \, d\rho \, d\varphi \, dz$.

Finally, it is possible to apply the final formula to cylindrical coordinates:

$$\iiint_D f(x, y, z)dx\,dy\,dz = \iiint_T f(\rho \cos \varphi, \rho \sin \varphi, z)$$

This method is convenient in case of cylindrical or conical domains or in regions where it is easy to individuate the z interval and even transform the circular base and the function.

Example: The function is $f(x, y, z) = x^2 + y^2 + z$ and as integration domain this cylinder: $D = \{x^2 + y^2 \leq 9, -5 \leq z \leq 5\,\}$. The transformation of D in cylindrical coordinates is the following:

$$T = \{0 \leq \rho \leq 3, 0 \leq \varphi \leq 2\pi, -5 \leq z \leq 5\}.$$

while the function becomes

$$f(\rho \cos \varphi, \rho \sin \varphi, z) = \rho^2 + z$$

Finally one can apply the integration formula:

$$\iiint_D \left(x^2 + y^2 + z\right) dx\, dy\, dz = \iiint_T \left(\rho^2 + z\right)\rho \, d\rho \, d\varphi \, dz;$$

developing the formula you have,

$$\int_{-5}^{5} dz \int_{0}^{2\pi} d\varphi \int_{0}^{3} \left(\rho^3 + \rho z\right) d\rho = 2\pi \int_{-5}^{5} \left[\frac{\rho^4}{4} + \frac{\rho^2 z}{2}\right]_0^3 dz = 2\pi \int_{-5}^{5} \left(\frac{81}{4} + \frac{9}{2}z\right) ddz = \cdots = 405\pi.$$

Spherical Coordinates

In R³ some domains have a spherical symmetry, so it's possible to specify the coordinates of every point of the integration region by two angles and one distance. It's possible to use therefore the *passage to spherical coordinates*; the function is transformed by this relation:

$$f(x, y, z) \rightarrow f(\rho \cos \theta \sin \varphi, \rho \sin \theta \sin \varphi, \rho \cos \varphi)$$

Points on the z-axis do not have a precise characterization in spherical coordinates, so θ can vary between 0 and 2π.

The better integration domain for this passage is the sphere.

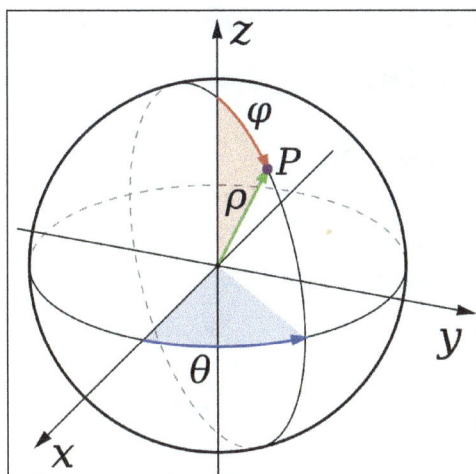

Spherical coordinates.

Example: The domain is $D = x^2 + y^2 + z^2 \leq 16$ (sphere with radius 4 and center at the origin); applying the transformation you get the region,

$$T = \{0 \leq \rho \leq 4, 0 \leq \varphi \leq \pi, 0 \leq \theta \leq 2\pi\}.$$

The Jacobian determinant of this transformation is the following:

$$\frac{\partial(x,y,z)}{\partial(\rho,\theta,\varphi)} = \begin{vmatrix} \cos\theta\sin\varphi & -\rho\sin\theta\sin\varphi & \rho\cos\theta\cos\varphi \\ \sin\theta\sin\varphi & \rho\cos\theta\sin\varphi & \rho\sin\theta\cos\varphi \\ \cos\varphi & 0 & -\rho\sin\varphi \end{vmatrix} = \rho^2\sin\varphi$$

The $dx\,dy\,dz$ differentials therefore are transformed to $\rho^2 \sin(\varphi)\,d\rho\,d\theta\,d\varphi$.

This yields the final integration formula:

$$\iiint_D f(x,y,z)\,dx\,dy\,dz = \iiint_T (\rho\sin\varphi\cos\theta, \rho\sin\varphi\sin\theta, \rho\cos\varphi)\rho^2\sin\varphi\,d\rho\,d\theta\,d\varphi.$$

It is better to use this method in case of spherical domains and in case of functions that can be easily simplified by the first fundamental relation of trigonometry extended to R³; in other cases it can be better to use cylindrical coordinates.

$$\iiint_T f(a,b,c)\rho^2\sin\varphi\,d\rho\,d\theta\,d\varphi.$$

The extra ρ^2 and $\sin\varphi$ come from the Jacobian.

In the following examples the roles of φ and θ have been reversed.

Example: D is the same region as in example above and $f(x, y, z) = x^2 + y^2 + z^2$ is the function to integrate. Its transformation is very easy:

$$f(\rho\sin\varphi\cos\theta, \rho\sin\varphi\sin\theta, \rho\cos\varphi) = \rho^2,$$

while we know the intervals of the transformed region T from D:

$$T = \{0 \le \rho \le 4, 0 \le \varphi \le \pi, 0 \le \theta \le 2\pi\}.$$

We therefore apply the integration formula:

$$\iiint_D \left(x^2 + y^2 + z^2\right) dx\, dy\, dz = \iiint_T \rho^2 \rho^2 \sin\theta\, d\rho\, d\theta\, d\varphi,$$

and, developing, we get

$$\iiint_T \rho^4 \sin\theta\, d\rho\, d\theta\, d\varphi = \int_0^\pi \sin\varphi\, d\varphi \int_0^4 \rho^4 d\rho \int_0^{2\pi} d\theta = 2\pi \int_0^\pi \sin\varphi \left[\frac{\rho^5}{5}\right]_0^4 d\varphi = 2\pi \left[\frac{\rho^5}{5}\right]_0^4 \left[-\cos\varphi\right]_0^\pi = \frac{4096\pi}{5}.$$

Example: The domain D is the ball with center at the origin and radius $3a$,

$$D = \left\{x^2 + y^2 + z^2 \le 9a^2\right\}$$

and $f(x, y, z) = x^2 + y^2$ is the function to integrate.

Looking at the domain, it seems convenient to adopt the passage to spherical coordinates, in fact, the intervals of the variables that delimit the new T region are obviously:

$$T = \{0 \le \rho \le 3a, 0 \le \varphi \le 2\pi, 0 \le \theta \le \pi\}.$$

However, applying the transformation, we get

$$f(x, y, z) = x^2 + y^2 \to \rho^2 \sin^2\theta \cos^2\varphi + \rho^2 \sin^2\theta \sin^2\varphi = \rho^2 \sin^2\theta.$$

Applying the formula for integration we obtain:

$$\iiint_T \rho^2 \sin^2\theta \rho^2 \sin\theta\, d\rho\, d\theta\, d\varphi = \iiint_T \rho^4 \sin^3\theta\, d\rho\, d\theta\, d\varphi$$

which is very hard to solve. This problem will be solved by using the passage to cylindrical coordinates. The new T intervals are

$$T = \left\{0 \le \rho \le 3a, 0 \le \varphi \le 2\pi, -\sqrt{9a^2 - \rho^2} \le z \le \sqrt{9a^2 - \rho^2}\right\};$$

the z interval has been obtained by dividing the ball into two hemispheres simply by solving the inequality from the formula of D (and then directly transforming $x^2 + y^2$ into ρ^2). The new function is simply ρ^2. Applying the integration formula:

$$\iiint_T \rho^2 \rho\, d\rho\, d\varphi\, dz.$$

Then we get

$$\int_0^{2\pi} d\varphi \int_0^{3a} \rho^3 d\rho \int_{-\sqrt{9a^2-\rho^2}}^{\sqrt{9a^2-\rho^2}} dz = 2\pi \int_0^{3a} 2\rho^3 \sqrt{9a^2 - \rho^2}\, d\rho$$

$$= -2\pi \int_{9a^2}^{0} (9a^2 - t)\sqrt{t}\, dt \qquad\qquad t = 9a^2 - \rho^2$$

$$= 2\pi \int_{0}^{9a^2} \left(9a^2\sqrt{t} - t\sqrt{t}\right) dt$$

$$= 2\pi \left(\int_{0}^{9a^2} 9a^2\sqrt{t}\, dt - \int_{0}^{9a^2} t\sqrt{t}\, dt \right)$$

$$= 2\pi \left[9a^2 \frac{2}{3} t^{\frac{3}{2}} - \frac{2}{5} t^{\frac{5}{2}} \right]_{0}^{9a^2}$$

$$= 2\cdot 27\pi a^5 \left(6 - \frac{18}{5} \right)$$

$$= \frac{648\pi}{5} a^5.$$

Thanks to the passage to cylindrical coordinates it was possible to reduce the triple integral to an easier one-variable integral.

Double Integral over a Rectangle

Let us assume that we wish to integrate a multivariable function f over a region A:

$$A = \left\{ (x, y) \in \mathbb{R}^2 : 11 \le x \le 14 \,;\, 7 \le y \le 10 \right\} \text{ and } f(x, y) = x^2 + 4y$$

From this we formulate the iterated integral,

$$\int_{7}^{10} \int_{11}^{14} (x^2 + 4y)\, dx\, dy$$

The inner integral is performed first, integrating with respect to x and taking y as a constant, as it is not the variable of integration. The result of this integral, which is a function depending only on y, is then integrated with respect to y.

$$\int_{11}^{14} \left(x^2 + 4y \right) dx = \left[\frac{1}{3} x^3 + 4yx \right]_{x=11}^{x=14}$$

$$= \frac{1}{3}(14)^3 + 4y(14) - \frac{1}{3}(11)^3 - 4y(11)$$

$$= 471 + 12y$$

We then integrate the result with respect to y.

$$\int_{7}^{10} (471 + 12y)\, dy = \left[471y + 6y^2 \right]_{y=7}^{y=10}$$

$$= 471(10) + 6(10)^2 - 471(7) - 6(7)^2$$

$$= 1719$$

In cases where the double integral of the absolute value of the function is finite, the order of integration is interchangeable, that is, integrating with respect to x first and integrating with respect to y first produce the same result. That is Fubini's theorem. For example, doing the previous calculation with order reversed gives the same result:

$$\int_{11}^{14}\int_{7}^{10}\left(x^2+4y\right) dy\, dx = \int_{11}^{14}\left[x^2 y+2y^2\right]_{y=7}^{y=10} dx$$
$$= \int_{11}^{14}(3x^2+102)\, dx$$
$$= \left[x^3+102x\right]_{x=11}^{x=14}$$
$$= 1719.$$

Double Integral over a Normal Domain

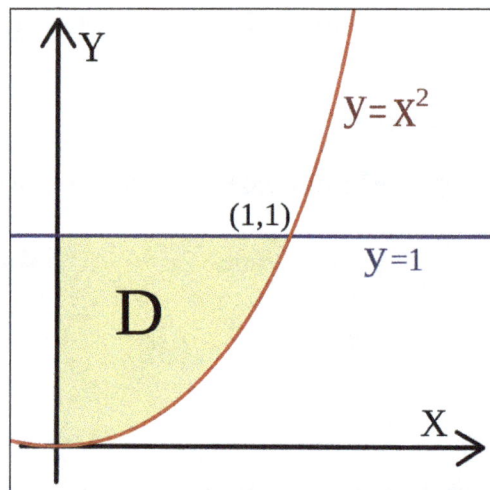

Example: double integral over the normal region D.

Consider the region:

$$D = \{(x,y)\in\mathrm{R}^2 : x\geq 0, y\leq 1, y\geq x^2\}$$

Calculate,

$$\iint_D (x+y)\,dx\,dy.$$

This domain is normal with respect to both the x- and y-axes. To apply the formulae it is required to find the functions that determine D and the intervals over which these functions are defined. In this case the two functions are:

$$\alpha(x) = x^2 \text{ and } \beta(x) = 1$$

while the interval is given by the intersections of the functions with $x = 0$, so the interval is $[a, b]$ = [0, 1] (normality has been chosen with respect to the x-axis for a better visual understanding).

It is now possible to apply the formula:

$$\iint_D (x+y)\,dx\,dy = \int_0^1 dx \int_{x^2}^1 (x+y)\,dy = \int_0^1 dx \left[xy + \frac{y^2}{2} \right]_{x^2}^1$$

(at first the second integral is calculated considering x as a constant). The remaining operations consist of applying the basic techniques of integration:

$$\int_0^1 \left[xy + \frac{y^2}{2} \right]_{x^2}^1 dx = \int_0^1 \left(x + \frac{1}{2} - x^3 - \frac{x^4}{2} \right) dx = \cdots = \frac{13}{20}.$$

If we choose normality with respect to the y-axis we could calculate

$$\int_0^1 dy \int_0^{\sqrt{y}} (x+y)\,dx.$$

and obtain the same value.

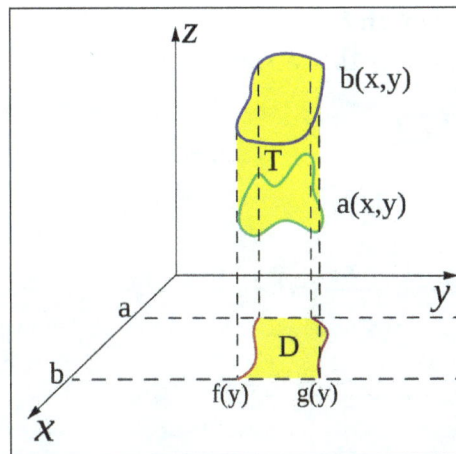

Example of domain in **R**³ that is normal with respect to the xy-plane.

Calculating Volume

Using the methods previously described, it is possible to calculate the volumes of some common solids.

- Cylinder: The volume of a cylinder with height h and circular base of radius R can be calculated by integrating the constant function h over the circular base, using polar coordinates.

$$\text{Volume} = \int_0^{2\pi} d\varphi \int_0^R h\rho\,d\rho = 2\pi h \left[\frac{\rho^2}{2} \right]_0^R = \pi R^2 h$$

This is in agreement with the formula for the volume of a prism

$$\text{Volume} = \text{base area} \times \text{height}.$$

- Sphere: The volume of a sphere with radius R can be calculated by integrating the constant function 1 over the sphere, using spherical coordinates.

$$\text{Volume} = \iiint_D f(x,y,z)\,dx\,dy\,dz$$
$$= \iiint_D 1\,dV$$
$$= \iiint_S \rho^2 \sin\varphi\,d\rho\,d\theta\,d\varphi$$
$$= \int_0^{2\pi} d\theta \int_0^{\pi} \sin\varphi\,d\varphi \int_0^R \rho^2\,d\rho$$
$$= 2\pi \int_0^{\pi} \sin\varphi\,d\varphi \int_0^R \rho^2\,d\rho$$
$$= 2\pi \int_0^{\pi} \sin\varphi \frac{R^3}{3}\,d\varphi$$
$$= \frac{2}{3}\pi R^3 \left[-\cos\varphi\right]_0^{\pi} = \frac{4}{3}\pi R^3.$$

- Tetrahedron (triangular pyramid or 3-simplex): The volume of a tetrahedron with its apex at the origin and edges of length ℓ along the x-, y- and z-axes can be calculated by integrating the constant function 1 over the tetrahedron.

$$\text{Volume} = \int_0^{\ell} dx \int_0^{\ell-x} dy \int_0^{\ell-x-y} dz$$
$$= \int_0^{\ell} dx \int_0^{\ell-x} (\ell-x-y)\,dy$$
$$= \int_0^{\ell} \left(l^2 - 2\ell x + x^2 - \frac{(\ell-x)^2}{2} \right) dx$$
$$= \ell^3 - \ell\ell^2 + \frac{\ell^3}{3} - \left[\frac{\ell^2 x}{2} - \frac{\ell x^2}{2} + \frac{x^3}{6} \right]_0^{\ell}$$
$$= \frac{\ell^3}{3} - \frac{\ell^3}{6} = \frac{\ell^3}{6}$$

This is in agreement with the formula for the volume of a pyramid

$$\text{Volume} = \frac{1}{3} \times \text{base area} \times \text{height} = \frac{1}{3} \times \frac{\ell^2}{2} \times \ell = \frac{\ell^3}{6}.$$

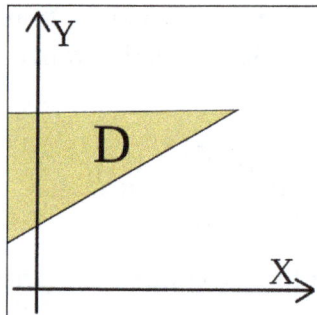

Example of an improper domain.

Multiple Improper Integral

In case of unbounded domains or functions not bounded near the boundary of the domain, we have to introduce the double improper integral or the triple improper integral.

Multiple Integrals and Iterated Integrals

Fubini's theorem states that if:

$$\iint_{A \times B} |f(x,y)| \, d(x,y) < \infty,$$

that is, if the integral is absolutely convergent, then the multiple integral will give the same result as either of the two iterated integrals:

$$\iint_{A \times B} f(x,y) \, d(x,y) = \int_A \left(\int_B f(x,y) \, dy \right) dx = \int_B \left(\int_A f(x,y) \, dx \right) dy.$$

In particular this will occur if $|f(x,y)|$ is a bounded function and A and B are bounded sets.

If the integral is not absolutely convergent, care is needed not to confuse the concepts of *multiple integral* and *iterated integral*, especially since the same notation is often used for either concept. The notation:

$$\int_0^1 \int_0^1 f(x,y) \, dy \, dx$$

means, in some cases, an iterated integral rather than a true double integral. In an iterated integral, the outer integral:

$$\int \quad dx$$

is the integral with respect to x of the following function of x:

$$g(x) = \int_0^1 f(x,y) \, dy.$$

A double integral, on the other hand, is defined with respect to area in the xy-plane. If the double integral exists, then it is equal to each of the two iterated integrals (either "$dy \, dx$" or "$dx \, dy$") and one often computes it by computing either of the iterated integrals. But sometimes the two iterated integrals exist when the double integral does not, and in some such cases the two iterated integrals are different numbers, i.e., one has,

$$\int_0^1 \int_0^1 f(x,y) \, dy \, dx \neq \int_0^1 \int_0^1 f(x,y) \, dx \, dy.$$

This is an instance of rearrangement of a conditionally convergent integral.

On the other hand, some conditions ensure that the two iterated integrals are equal even though the double integral need not exist. By the Fichtenholz–Lichtenstein theorem, if f is bounded on $[0, 1] \times [0, 1]$ and both iterated integrals exist, then they are equal. Moreover, existence of the inner integrals ensures existence of the outer integrals. The double integral need not exist in this case even as Lebesgue integral, according to Sierpiński.

The notation,

$$\int_{[0,1]\times[0,1]} f(x,y)\,dx\,dy$$

may be used if one wishes to be emphatic about intending a double integral rather than an iterated integral.

Practical Applications

Quite generally, just as in one variable, one can use the multiple integral to find the average of a function over a given set. Given a set $D \subseteq \mathrm{R}^n$ and an integrable function f over D, the average value of f over its domain is given by,

$$\bar{f} = \frac{1}{m(D)} \int_D f(x)\,dx,$$

where $m(D)$ is the measure of D.

Additionally, multiple integrals are used in many applications in physics.

In mechanics, the moment of inertia is calculated as the volume integral (triple integral) of the density weighed with the square of the distance from the axis:

$$I_z = \iiint_V \rho r^2\,dV.$$

The gravitational potential associated with a mass distribution given by a mass measure dm on three-dimensional Euclidean space R^3 is,

$$V(\mathrm{x}) = -\iiint_{\mathrm{R}^3} \frac{G}{|\,\mathrm{x}\text{-}\mathrm{y}\,|}\,dm(\mathrm{y}).$$

If there is a continuous function $\rho(\mathrm{x})$ representing the density of the distribution at x, so that $dm(\mathrm{x}) = \rho(\mathrm{x})d^3\,\mathrm{x}$, where $d^3\mathrm{x}$ is the Euclidean volume element, then the gravitational potential is,

$$V(\mathrm{x}) = -\iiint_{\mathrm{R}^3} \frac{G}{|\,\mathrm{x}\text{-}\mathrm{y}\,|}\rho(\mathrm{y})d^3\,\mathrm{y}.$$

In electromagnetism, Maxwell's equations can be written using multiple integrals to calculate the total magnetic and electric fields. In the following example, the electric field produced by a distribution of charges given by the volume charge density $\rho(\,\vec{r}\,)$ is obtained by a *triple integral* of a vector function:

$$\vec{E} = \frac{1}{4\pi\varepsilon_0} \iiint \frac{\vec{r} - \vec{r}'}{\|\vec{r} - \vec{r}'\|^3}\rho(\vec{r}')d^3 r'.$$

This can also be written as an integral with respect to a signed measure representing the charge distribution.

MATRIX CALCULUS

Matrix calculus is concerned with rules for operating on functions of matrices. For example, suppose that an $m \times n$ matrix X is mapped into a $p \times q$ matrix Y. We are interested in obtaining expressions for derivatives such as,

$$\frac{\partial Y_{ij}}{\partial X_{kl}},$$

for all i, j and k, l. The main difficulty here is keeping track of where things are put. There is no reason to use subscripts; it is far better instead to use a system for ordering the results using matrix operations.

Matrix calculus makes heavy use of the vec operator and Kronecker products. The vec operator vectorizes a matrix by stacking its columns (it is convention that column rather than row stacking is used). For example, vectorizing the matrix:

$$\begin{bmatrix} 1 & 2 \\ 3 & 4 \\ 5 & 6 \end{bmatrix}$$

produces

$$\begin{bmatrix} 1 \\ 3 \\ 5 \\ 2 \\ 4 \\ 6 \end{bmatrix}$$

The Kronecker product of two matrices, A and B, where A is $m \times n$ and B is $p \times q$, is defined as:

$$A \otimes B = \begin{bmatrix} A_{11}B & A_{12}B & \dots & A_{1n}B \\ A_{21}B & A_{22}B & \dots & A_{2n}B \\ \dots & \dots & \dots & \dots \\ A_{m1}B & A_{m2}B & \dots & A_{mn}B \end{bmatrix}$$

which is an $mp \times nq$ matrix. There is an important relationship between the Kronecker product and the vec operator:

$$\text{vec}(AXB) = (B^\top > \otimes A)\text{vec}(X).$$

This relationship is extremely useful in deriving matrix calculus results.

Another matrix operator that will prove useful is one related to the vec operator. Define the matrix $T_{m,n}$ as the matrix that transforms vec (A) into $\text{vec}(A^\top)$:

$$T_{m,n}\,\text{vec}(A) = \text{vec}(A^\top).$$

Note the size of this matrix is $mn \times mn$. $T_{m,n}$ has a number of special properties. The first is clear from its definition; if $T_{m,n}$ is applied to the vec of an m × n matrix and then $T_{n,m}$ applied to the result, the original vectorized matrix results:

$$T_{n,m}T_{m,n}\,\text{vec}(A) = \text{vec}(A)$$

thus,

$$T_{n,m}T_{m,n} = I_{mn}.$$

The fact that,

$$T_{n,m} = T_{m,n}^{-1}$$

follows directly. Perhaps less obvious is that,

$$T_{m,n} = T_{n,m}^\top$$

(also combining these results means that $T_{m,n}$ is an orthogonal matrix).

The matrix operator $T_{m,n}$ is a permutation matrix, i.e., it is composed of 0s and 1s, with a single 1 on each row and column. When premultiplying another matrix, it simply rearranges the ordering of rows of that matrix (postmultiplying by $T_{m,n}$ rearranges columns).

The transpose matrix is also related to the Kronecker product. With A and B defined as above,

$$B \otimes A = T_{p,m}(A \otimes B)T_{n,q}.$$

This can be shown by introducing an arbitrary $n \times q$ matrix C:

$$T_{p,m}(A \otimes B)T_{n,q}\,\text{vec}(C) = T_{p,m}(A \otimes B)\text{vec}(C^\top)$$
$$= T_{p,m}\,\text{vec}(BC^\top A^\top)$$
$$= \text{vec}(ACB^\top)$$
$$= (B \otimes A)\text{vec}(C).$$

This implies that $\left((B \otimes A) - T_{p,m} (A \otimes B) T_{n,q} \right) \text{vec}(C) = 0$. Because C is arbitrary, the desired result must hold.

An immediate corollary to the above result is that,

$$(A \otimes B) T_{n,q} = T_{m,p} (B \otimes A).$$

It is also useful to note that $T_{1,m} = T_{m,1} = I_m$. Thus, if A is $1 \times n$ then $(A \otimes B) T_{n,q} = (B \otimes A)$. When working with derivatives of scalars this can result in considerable simplification.

Turning now to calculus, define the derivative of a function mapping $\Re^n \to \Re^m$ as the $m \times n$ matrix of partial derivatives:

$$[Df]_{ij} = \frac{\partial f_i(x)}{\partial x_j}.$$

For example, the simplest derivative is,

$$\frac{dAx}{dx} = A.$$

Using this definition, the usual rules for manipulating derivatives apply naturally if one respects the rules of matrix conformability. The summation rule is obvious:

$$D\left[\alpha f(x) + \beta g(x) \right] = \alpha Df(x) + \beta Dg(x),$$

where α and β are scalars. The chain rule involves matrix multiplication, which requires conformability. Given two functions $f : \Re^n \to \Re^m$ and $g : \Re^n \to \Re^m$, the derivative of the composite function is,

$$D\left[f(g(x)) \right] = f'(g(x)) g'(x).$$

Notice that this satisfies matrix multiplication conformability, whereas the expression $g'(x) f'(g(x))$ attempts to postmultipy an $n \times p$ matrix by an $m \times n$ matrix. To define a product rule, consider the expression $f(x)^{\mathsf{T}} g(x)$, where $f, g : \Re^n \to \Re^m$. The derivative is the $1 \times n$ vector given by,

$$D\left[f(x)^{\mathsf{T}} g(x) \right] = g(x)^{\mathsf{T}} f'(x) + f(x)^{\mathsf{T}} g'(x).$$

Notice that no other way of multiplying g by f' and f by g' would ensure conformability. A more general version of the product rule is defined below.

The product rule leads to a useful result about quadratic functions:

$$\frac{dx^{\mathsf{T}} Ax}{dx} = x^{\mathsf{T}} A + x^{\mathsf{T}} A^{\mathsf{T}} = x^{\mathsf{T}} \left(A + A^{\mathsf{T}} \right).$$

When A is symmetric this has the very natural form $dx^{\mathsf{T}} Ax / dx = 2x^{\mathsf{T}} A$.

These rules define derivatives for vectors. Defining derivatives of matrices with respect to matrices is accomplished by vectorizing the matrices, so $dA(X)/dX$ is the same thing as $d\mathrm{vec}(A(X))/d\mathrm{vec}(X)$. This is where the the relationship between the vec operator and Kronecker products is useful. Consider differentiating $dx^\mathsf{T}Ax$ with respect to A (rather than with respect to x as above):

$$\frac{d\mathrm{vec}(x^\mathsf{T}Ax)}{d\mathrm{vec}(A)} = \frac{d(x^\mathsf{T}\otimes x^\mathsf{T})\mathrm{vec}(A)}{d\mathrm{vec}(A)} = (x^\mathsf{T}\otimes x^\mathsf{T})$$

(the derivative of an $m\times n$ matrix A with respect to itself is I_{mn}).

A more general product rule can be defined. Suppose that $f:\mathfrak{R}^n\to\mathfrak{R}^{m\times p}$ and $g:\mathfrak{R}^n\to\mathfrak{R}^{p\times q}$, so $f(x)g(x):\mathfrak{R}^n\to\mathfrak{R}^{m\times q}$. Using the relationship between the vec and Kronecker product operators

$$\mathrm{vec}(I_m f(x)g(x)I_q) = (g(x)^\mathsf{T}\otimes I_m)\mathrm{vec}(f(x)) = (I_q\otimes f(x))\mathrm{vec}(g(x)).$$

A natural product rule is therefore,

$$Df(x)g(x) = (g(x)^\mathsf{T}\otimes I_m)f'(x) + (I_q\otimes f(x))g'(x).$$

This can be used to determine the derivative of $dA^\mathsf{T}A/dA$ where A is m × n.

$$\mathrm{vec}(A^\mathsf{T}A) = (I_n\otimes A^\mathsf{T})\mathrm{vec}(A) = (A^\mathsf{T}\otimes I_n)\mathrm{vec}(A^\mathsf{T}) = (A^\mathsf{T}\otimes I_n)T_{m,n}\mathrm{vec}(A).$$

Thus (using the product rule),

$$\frac{dA^\mathsf{T}A}{dA} = (I_n\otimes A^\mathsf{T}) + (A^\mathsf{T}\otimes I_n)T_{m,n}.$$

This can be simplified somewhat by noting that,

$$(A^\mathsf{T}\otimes I_n)T_{m,n} = T_{n,n}(I_n\otimes A^\mathsf{T}).$$

Thus,

$$\frac{dA^\mathsf{T}A}{dA} = (I_{n2} + T_{n,n})(I_n\otimes A^\mathsf{T}).$$

The product rule is also useful in determining the derivative of a matrix inverse:

$$\frac{dA^{-1}A}{dA} = (A^\mathsf{T}\otimes I_n)\frac{dA^{-1}}{dA} + (I_n\otimes A^{-1}).$$

But $A^{-1}A$ is identically equal to I, so its derivative is identically o. Thus,

$$\frac{dA^{-1}}{dA} = -(A^\mathsf{T}\otimes I_n)^{-1}(I_n\otimes A^{-1}) = -(A^{-\mathsf{T}}\otimes I_n)(I_n\otimes A^{-1}) = -(A^{-\mathsf{T}}\otimes A^{-1}).$$

It is also useful to have an expression for the derivative of a determinant. Suppose A is $n \times n$ with $|A| \neq 0$. The determinant can be written as the product of the ith row of the adjoint of A (A^*) with the ith column of A:

$$|A| = A^*_{i.} A_{.i} \; .$$

Recall that the elements of the ith row of A^* are not influenced by the elements in the ith column of A and hence,

$$\frac{\partial |A|}{\partial A_{.i}} = A^*_{i.}.$$

To obtain the derivative with respect to all of the elements of A, we can concatenate the partial derivatives with respect to each column of A:

$$\frac{d|A|}{dA} = \begin{bmatrix} A^*_{1.} & A^*_{2.} & \ldots & A^*_{n.} \end{bmatrix} = |A| \begin{bmatrix} [A^{-1}]_{1.} & [A^{-1}]_{2.} & \ldots & [A^{-1}]_{n.} \end{bmatrix} = |A| \, \mathrm{vec}\left(A^{-\mathrm{T}}\right)^{\mathrm{T}}.$$

The following result is an immediate consequence,

$$\frac{d \ln |A|}{dA} = \mathrm{vec}\left(A^{-\mathrm{T}}\right)^{\mathrm{T}}.$$

Matrix differentiation results allow us to compute the derivatives of the solutions to certain classes of equilibrium problems. Consider, for example, the solution, x, to a linear complementarity problem LCP(M,q) that solves,

$$Mx + q \geq 0, \quad x \geq 0, \quad x^{\mathrm{T}}\left(Mx + q\right) = 0$$

The ith element of x is either exactly equal to 0 or is equal to the ith element of $Mx + q$. Define a diagonal matrix D such that $D_{ii} = 1$ if $x > 0$ and equal 0 otherwise. The solution can then be written as $x = -\hat{M}^{-1} Dq$, where $\hat{M} = DM + I - D$. It follows that,

$$\frac{\partial x}{\partial q} = -\hat{M}^{-1} D$$

and that,

$$\begin{aligned}
\frac{\partial x}{\partial M} &= \frac{\partial x}{\partial \hat{M}^{-1}} \frac{\partial \hat{M}^{-1}}{\partial \hat{M}} \frac{\partial \hat{M}}{\partial M} \\
&= \left(-q^{\mathrm{T}} D \otimes I\right)\left(-\hat{M}^{-\mathrm{T}} \otimes \hat{M}^{-1}\right)\left(I \otimes D\right) \\
&= q > D\hat{M}^{-\mathrm{T}} > \otimes \hat{M}^{-1} D \\
&= x^{\mathrm{T}} \otimes \partial x / \partial q
\end{aligned}$$

Given the prevalence of Kronecker products in matrix derivatives, it would be useful to have rules for computing derivatives of Kronecker products themselves, i.e. $dA \otimes B / dA$ and $dA \otimes B / dB$. Because each element of a Kronecker product involves the product of one element from A multiplied by one element of B, the derivative $dA \otimes B/dA$ must be composed of zeros and the elements of B arranged in a certain fashion. Similarly, the derivative $dA \otimes B / dB$ is composed of zeros and the elements of A arranged in a certain fashion.

It can be verified that $dA \otimes B / dA$ can be written as:

$$\frac{dA \otimes B}{dA} = \begin{bmatrix} \Psi_1 & 0 & \cdots & 0 \\ \Psi_2 & 0 & \cdots & 0 \\ \cdots & \cdots & \cdots & \cdots \\ \Psi_q & 0 & \cdots & 0 \\ 0 & \Psi_1 & \cdots & 0 \\ 0 & \Psi_2 & \cdots & 0 \\ \cdots & \cdots & \cdots & \cdots \\ 0 & \Psi_q & \cdots & 0 \\ \cdots & \cdots & \cdots & \cdots \\ 0 & 0 & \cdots & \Psi_1 \\ 0 & 0 & \cdots & \Psi_2 \\ \cdots & \cdots & \cdots & \cdots \\ 0 & 0 & \cdots & \Psi_q \end{bmatrix} = I_n \otimes \begin{bmatrix} \Psi_1 \\ \Psi_2 \\ \cdots \\ \Psi_q \end{bmatrix}$$

where,

$$\Psi_i = \begin{bmatrix} B_{1i} & 0 & \cdots & 0 \\ B_{2i} & 0 & \cdots & 0 \\ \cdots & \cdots & \cdots & \cdots \\ B_{pi} & 0 & \cdots & 0 \\ 0 & B_{1i} & \cdots & 0 \\ 0 & B_{2i} & \cdots & 0 \\ \cdots & \cdots & \cdots & \cdots \\ 0 & B_{pi} & \cdots & 0 \\ \cdots & \cdots & \cdots & \cdots \\ 0 & 0 & \cdots & B_{1i} \\ 0 & 0 & \cdots & B_{2i} \\ \cdots & \cdots & \cdots & \cdots \\ 0 & 0 & \cdots & B_{pi} \end{bmatrix} = I_m \otimes B_{\cdot i}$$

This can be written more compactly as,

$$\frac{dA \otimes B}{dA} = \left(I_n \otimes T_{qm} \otimes I_p\right)\left(I_{mn} \otimes \text{vec}(B)\right) = \left(I_{nq} \otimes T_{mp}\right)\left(I_n \otimes \text{vec}(B) \otimes I_m\right).$$

Similarly $dA \otimes B / dB$ can be written as:

$$\frac{dA \otimes B}{dB} = \begin{bmatrix} \Theta_1 & 0 & \cdots & 0 \\ 0 & \Theta_1 & \cdots & 0 \\ \cdots & \cdots & \cdots & \cdots \\ 0 & 0 & \cdots & \Theta_1 \\ \Theta_2 & 0 & \cdots & 0 \\ 0 & \Theta_2 & \cdots & 0 \\ \cdots & \cdots & \cdots & \cdots \\ 0 & 0 & \cdots & \Theta_2 \\ \cdots & \cdots & \cdots & \cdots \\ \Theta_q & 0 & \cdots & 0 \\ 0 & \Theta_q & \cdots & 0 \\ \cdots & \cdots & \cdots & \cdots \\ 0 & 0 & \cdots & \Theta_q \end{bmatrix}$$

where,

$$\Theta_i = \begin{bmatrix} A_{i1} & 0 & \cdots & 0 \\ 0 & A_{i1} & \cdots & 0 \\ \cdots & \cdots & \cdots & \cdots \\ 0 & 0 & \cdots & A_{i1} \\ A_{i2} & 0 & \cdots & 0 \\ 0 & A_{i2} & \cdots & 0 \\ \cdots & \cdots & \cdots & \cdots \\ 0 & 0 & \cdots & A_{i2} \\ \cdots & \cdots & \cdots & \cdots \\ A_{im} & 0 & \cdots & 0 \\ 0 & A_{im} & \cdots & 0 \\ \cdots & \cdots & \cdots & \cdots \\ 0 & 0 & \cdots & A_{im} \end{bmatrix}$$

This can be written more compactly as,

$$\frac{dA \otimes B}{dB} = \left(I_n \otimes T_{qm} \otimes I_p\right)\left(vec(A) \otimes I_{pq}\right) = \left(T_{pq} \otimes I_{mn}\right)\left(I_q \otimes vec(A) \otimes I_p\right).$$

Notice that if either A is a row vector $(m = 1)$ or B is a column vector $(q = 1)$, the matrix $\left(I_n \otimes T_{qm} \otimes I_p\right) = Imnpq$ and hence can be ignored.

To illustrate a use for these relationships, consider the second derivative of $xx^\top$ with respect to x, an n-vector.

$$\frac{dxx^\top}{dx} = x \otimes I_n + I_n \otimes x;$$

hence,

$$\frac{d^2 xx^\top}{dx dx^\top} = \left(T_{nn} \otimes I_n\right)\left(I_n \otimes \mathrm{vec}\left(I_n\right) + \mathrm{vec}\left(I_n\right) \otimes I_n\right).$$

Another example is,

$$\frac{d^2 A^{-1}}{dA dA} = \left(I_n \otimes T_{nn} \otimes I_n\right)\left[\left(I_{n^2} \otimes \mathrm{vec}\left(A^{-1}\right)\right) T_{nn} + \left(\mathrm{vec}\,\left(A^{-\top}\right) \otimes I_{n^2}\right)\right]\left(A^{-\top} \otimes A^{-1}\right)$$

Often, especially in statistical applications, one encounters matrices that are symmetrical. It would not make sense to take a derivative with respect to the i, jth element of a symmetric matrix while holding the j, ith element constant. Generally it is preferable to work with a vectorized version of a symmetric matrix that excludes with the upper or lower portion of the matrix. The vech operator is typically taken to be the column-wise vectorization with the upper portion excluded:

$$\mathrm{vech}\left(A\right) = \begin{bmatrix} A_{11} \\ \cdots \\ A_{n,1} \\ A_{22} \\ \cdots \\ A_{n2} \\ \cdots \\ A_{nn-1} \\ A_{nn} \end{bmatrix}$$

One obtains this by selecting elements of vec(A) and therefore we can write:

$$\mathrm{vech}\left(A\right) = S_n \mathrm{vec}\left(A\right),$$

where S_n is an $n(n+1)/2 \times n^2$ matrix of 0s and 1s, with a single 1 in each row. The vech operator can be applied to lower triangular matrices as well; there is no reason to take derivatives with respect to the upper part of a lower triangular matrix (it can also be applied to the transpose of an upper triangular matrix). The use of the vech operator is also important in efficient computer storage of symmetric and triangular matrices.

To illustrate the use of the vech operator in matrix calculus applications, consider an $n \times n$ symmetric matrix C defined in terms of a lower triangular matrix, L,

$$C = LL^\mathsf{T}.$$

Using the already familiar methods it is easy to see that

$$\frac{dC}{dL} = (I_n \otimes L) T_{n,n} + (L \otimes I_n).$$

Using the chain rule

$$\frac{d\mathrm{vech}(C)}{d\mathrm{vech}(L)} = \frac{d\mathrm{vech}(C)}{dC} \frac{dC}{dL} \frac{dL}{d\mathrm{vech}(L)} = S_n \frac{dC}{dL} S_n^\mathsf{T}.$$

Inverting this expression provides an expression for dvech(L)/dvech(C).

Related to matrix derivatives is the issue of Taylor expansions of matrixto-matrix functions. One way to think of matrix derivatives is in terms of multidimensional arrays. An $mn \times pq$ matrix can also be thought of as an $m \times n \times p \times q$ 4-dimensional array. The "reshape" function in MATLAB implements such transformations. The ordering of the individual elements has not change, only the way the elements are indexed.

The dth order Taylor expansion of a function $f(X): R^{m \times n} \to R^{p \times q}$ at $\tilde{X}$ can be computed in the following way:

$$f = \mathrm{vec}\left(f(\tilde{X})\right)$$

$$dX = \mathrm{vec}\left(X - \tilde{X}\right)$$

for $i = 1$ to d

$\{$

$$f_i = f^{(i)}(\tilde{X}) dX$$

for $j = 2$ to i

$$f_i = reshape\left(f_i, mn(pq)^{j-2}, pq\right) dX / j$$

$$f = f + f_i$$

$\}$

$$f = reshape(f, m, n)$$

The techniques described thus far can be applied to computation of derivatives of common "special" functions. First, consider the derivative of a nonnegative integer power A^i of a square matrix A. Application of the chain rule leads to the recursive definition.

$$\frac{dA^i}{dA} = \frac{dA^{i-1}A}{dA} = \left(A^\mathsf{T} \otimes I\right)\frac{dA^{i-1}}{dA} + \left(I \otimes A^{i-1}\right)$$

which can also be expressed as a sum of i terms,

$$\frac{dA^i}{dA} = \sum_{j=1}^{i}(A^\mathsf{T})^{i-j} \otimes A^{j-1}.$$

This result can be used to derive an expression for the derivative of the matrix exponential function, which is defined in the usual way in terms of a Taylor expansion:

$$\exp(A) = \sum_{i=0}^{\infty}\frac{Ai}{i!}.$$

Thus,

$$\frac{d\exp(A)}{dA} = \sum_{i=0}^{\infty}\frac{1}{i!}\sum_{j=1}^{i}(A^\mathsf{T})^{i-j} \otimes A^{j-1}.$$

The same approach can be applied to the matrix natural logarithm:

$$\ln(A) = -\sum_{i=0}^{\infty}\frac{1}{i!}(I - A)^i.$$

Operator Results

A is $m \times n$, B is $p \times q$, X is defined comformably.

$$A \otimes B = \begin{bmatrix} A_{11}B & A_{12}B & \dots & A_{1n}B \\ A_{21}B & A_{22}B & \dots & A_{2n}B \\ \dots & \dots & \dots & \dots \\ A_{m1}B & A_{m2}B & \dots & A_{mn}B \end{bmatrix}$$

$$(AC \otimes BD) = (A \otimes B)(C \otimes D)$$

$$(A \otimes B)^{-1} = A^{-1} \otimes B^{-1}$$

$$(A \otimes B)^\mathsf{T} = A^\mathsf{T} \otimes B^\mathsf{T}$$

$$\mathrm{vec}(AXB) = (B^\mathsf{T} \otimes A)\mathrm{vec}(X)$$

$$\mathrm{trace}(AX) = \mathrm{vec}(A^\mathsf{T})^\mathsf{T}\mathrm{vec}(X)$$

$$\mathrm{trace}(AX) = \mathrm{trace}(XA)$$

$$T_{m,n}\mathrm{vec}(A) = \mathrm{vec}(A^\mathsf{T})$$

$$T_{n,m}T_{m,n} = I_{mn}$$

$$T_{n,m} = T_{m,n}^{-1}$$

$$T_{m,n} = T_{n,m}^{\mathsf{T}}$$

$$B \otimes A = T_{p,m}(A \otimes B)T_{n,q}$$

Differentiation Results

A is $m \times n$, B is $p \times q$, x is $n \times 1$, X is defined comformably.

$$[Df]_{ij} = \frac{df_i(x)}{dx_j}$$

$$\frac{dAx}{dx} = A$$

$$D\big[\alpha f(x) + \beta g(x)\big] = \alpha Df(x) + \beta Dg(x)$$

$$D\big[f(g(x))\big] = f'(g(x))g'(x)$$

$$D\big[f(x)g(x)\big] = \big(g(x)^{\mathsf{T}} \otimes I_m\big)f'(x) + \big(I_p \otimes f(x)\big)g'(x).$$

$$\frac{dx^{\mathsf{T}}Ax}{dx} = x^{\mathsf{T}}\big(A + A^{\mathsf{T}}\big)$$

$$\frac{d\mathrm{vec}\big(x^{\mathsf{T}}Ax\big)}{d\mathrm{vec}(A)} = \big(x^{\mathsf{T}} \otimes x^{\mathsf{T}}\big)$$

$$\frac{dA^{\mathsf{T}}A}{dA} = \big(I_{n^2} + T_{n,n}\big)\big(I_n \otimes A^{\mathsf{T}}\big)$$

$$\frac{dAA^{\mathsf{T}}}{dA} = \big(I_{m^2} + T_{m,m}\big)\big(A \otimes I_m\big)$$

$$\frac{dx^{\mathsf{T}}A^{\mathsf{T}}Ax}{dA} = 2x^{\mathsf{T}} \otimes x^{\mathsf{T}}A^{\mathsf{T}} \ \big(\text{when } x \text{ is a vector}\big)$$

$$\frac{dx^{\mathsf{T}}AA^{\mathsf{T}}x}{dA} = 2x^{\mathsf{T}}A \otimes x^{\mathsf{T}} \big(\text{when } x \text{ is a vector}\big)$$

$$\frac{dAXB}{dX} = B^{\mathsf{T}} \otimes A$$

$$\frac{dA^{-1}}{dA} = -\left(A^{-\mathsf{T}} \otimes A^{-1}\right)$$

$$\frac{d\ln|A|}{dA} = \mathrm{vec}\left(A^{-\mathsf{T}}\right)^{\mathsf{T}}$$

$$\frac{d\,\mathrm{trace}(AX)}{dX} = \mathrm{vec}\left(A^{\mathsf{T}}\right)^{\mathsf{T}}$$

$$\frac{dA \otimes B}{dA} = \left(I_n \otimes T_{qm} \otimes I_p\right)\left(I_{mn} \otimes vec(B)\right) = \left(I_{nq} \otimes T_{mp}\right)\left(I_n \otimes \mathrm{vec}(B) \otimes I_m\right).$$

$$\frac{dA \otimes B}{dB} = \left(I_n \otimes T_{qm} \otimes I_p\right)\left(\mathrm{vec}(A) \otimes I_{pq}\right) = \left(T_{pq} \otimes I_{mn}\right)\left(I_q \otimes \mathrm{vec}(A) \otimes I_p\right).$$

$$\frac{dxx^{\mathsf{T}}}{dx} = x \otimes I_n + I_n \otimes x$$

$$\frac{d^2 xx^{\mathsf{T}}}{dx\,dx^{\mathsf{T}}} = \left(T_{nn} \otimes I_n\right)\left(I_n \otimes \mathrm{vec}(I_n) + \mathrm{vec}(I_n) \otimes I_n\right)$$

$$\frac{dA^i}{dA} = \left(A^{\mathsf{T}} \otimes I\right)\frac{dA^{i-1}}{dA} + \left(I \otimes A^{i-1}\right) = \sum_{j=1}^{i}\left(A^{\mathsf{T}}\right)^{i-j} \otimes A^{j-1}.$$

$$\frac{d\exp(A)}{dA} = \sum_{i=0}^{\infty}\frac{1}{i!}\sum_{j=1}^{i}\left(A^{\mathsf{T}}\right)^{i-j} \otimes A^{j-1}.$$

References

- Rudin, Walter. Principles of Mathematical Analysis. Walter Rudin Student Series in Advanced Mathematics (3rd ed.). McGraw–Hill. ISBN 978-0-07-054235-8

- Multivariable-calculus, bytes: toppr.com, Retrieved 25 June, 2019

- Sinclair, George Edward (1974). "A finitely additive generalization of the Fichtenholz–Lichtenstein theorem". Transactions of the American Mathematical Society. AMS. 193: 359–374. doi:10.2307/1996919. JSTOR 1996919

- Matrixc1, resources, shared: 2.stat.duke.edu, Retrieved 11 April, 2019

Vector Calculus

The branch of calculus which deals with differentiation and integration of vector fields is known as vector calculus. Some of the differential operators studied within vector calculus are gradient, divergence and curl. All these diverse concepts related to vector calculus have been carefully analyzed in this chapter.

Vector Calculus is a mathematical discipline that studies the properties of operations on vectors of Euclidean space. the concept of a vector constitutes the mathematical abstraction of quantities that are characterized not only by a numerical value but also by a direction (for example, force, acceleration, velocity).

Vector Algebra

A vector is a directed line segment, that is, a segment whose beginning (also called the vector's point of application) and end are indicated. The length of the directed line segment, which represents a vector, is called its length or magnitude. The length of vector a is denoted by |a|. Vectors are called collinear if they lie either on the same line or on parallel lines. Two vectors are said to be equal if they are collinear and have the same length and direction. All zero vectors are considered to be equal. Vector calculus, deals with free vectors.

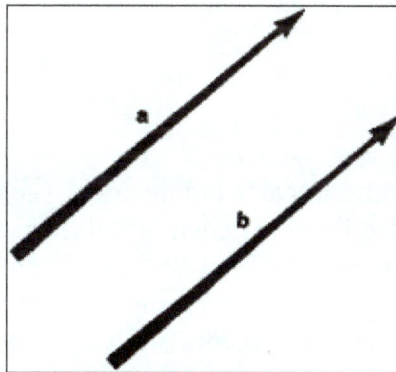

An important role in vector algebra is played by linear operations on vectors: adding vectors and multiplying them by real numbers. The sum a + b of vectors a and b is the vector extending from the beginning of vector a to the end of vector b such that the beginning of vector b is joined to the end of vector a. The derivation of this rule is related to the parallelogram rule of vector addition, whose source is the experimental fact of the addition of forces (vector magnitudes) according to this rule. The construction of the sum of several vectors is clear from figure. The product α a of vector a and the number α is a vector that is collinear with vector a and has a length |α| · |a| and a direction that coincides with the direction of a when α > 0 and is opposite to that of a when α < 0. Vector -1 · a is the inverse vector of a and is denoted by -a.

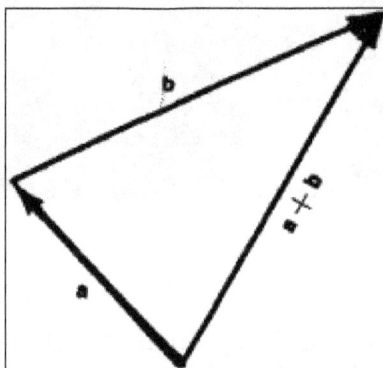

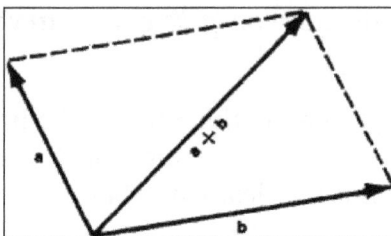

- $a + b = b + a$

- $(a + b) + c = a + (b + c)$

- $a + o = a$

- $a + (-a) = o$

- $1 \cdot a = a$

- $\alpha(\beta a) = (\alpha\beta)a$

- $\alpha(a + b) = \alpha a + \alpha b$

- $(\alpha + \beta)a = \alpha a + \beta a$

The concept of linearly dependent and linearly independent vectors is often encountered in vector algebra. Vectors $a_1, a_2,..., a_n$ are called linearly dependent if there exist such numbers $\alpha_1, \alpha_2,..., \alpha_n$, of which at least one of them differs from zero, that the linear combination $\alpha_1 a_1 + \cdots + \alpha_n a_n$ of these vectors is equal to zero. Vectors $a_1, a_2, ... , a_n$ that are not linearly dependent are called linearly independent. Let us note that any three nonzero vectors not lying in one plane are linearly independent.

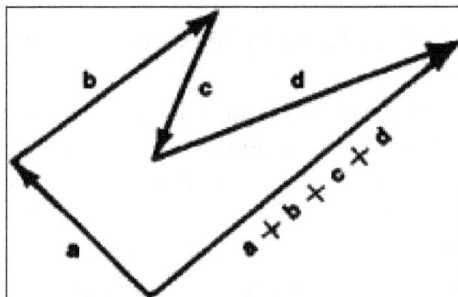

The vectors of Euclidean space have the following property: there exist three linearly independent vectors, and any arbitrary four vectors are linearly dependent. This property characterizes the three-dimensionality of the set of vectors under consideration. In conjunction with the properties listed above, the indicated property implies that the set of all vectors of Euclidean space forms a so-called vector space. The linearly independent vectors e_1, e_2, and e_3 form a basis. Any vector a can be uniquely resolved in terms of basis vectors: a = X e_1 + Y e_2 + Z e_3 the coefficients X, Y, and Z are called the coordinates (components) of vector a in the given basis. If vector a has coordinates X, Y, and Z, this can be written as a = {X, Y, Z}. Three mutually orthogonal (perpendicular) vectors, whose lengths are equal to one and which are usually denoted by i, j, and k, form a so-called orthonormalized basis. If these vectors are located with their initial points at one point O, they form a rectangular Cartesian coordinate system in space. The coordinates X, Y, Z of any point M in this system are defined as the coordinates of the vector O M. The linear operations on vectors, indicated previously, correspond to analogous operations on their coordinates: if the coordinates of vectors a and b are {X_1, Y_1, Z_1} and X_2, Y_2, Z_2} respectively, then the coordinates of the sum a + b of these vectors are {$X_1 + X_2$, $Y_1 + Y_2$, $Z_1 + Z_2$} the coordinates of vector λa are {λX_1, λY_1, λZ_1}.

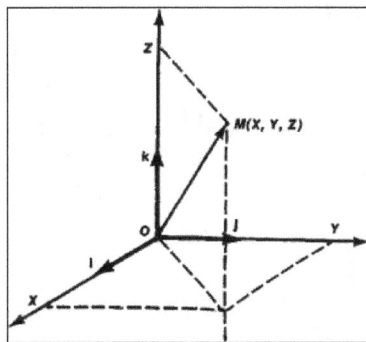

The development and application of vector algebra is closely connected with various types of products of vectors: scalar, vector, and mixed. The concept of the scalar product of vectors arises, for example, in examining the work performed by a force F along a given path S: the work is equal to |F| |S| cos φ, where φ is the angle between vectors F and S. Mathematically, the scalar product of vectors a and b is defined as the number denoted by *(a, b)* and equal to the product of the magnitudes of these vectors and of the cosine of the angle between them:

(a, b) = |a| |bx| cos φ

The quantity |b| cos φ is called the projection of vector b on the axis determined by vector a and is denoted by proj$_a$b. Therefore, (a, b) = | a | proj$_a$b. In particular, if a is a unit vector (|a| = 1), then (a, b) = proj$_a$b. The following properties of the scalar product are obvious:

(a, b) = (b, a)

(λa,b) = φ(a,b)

(a + b,c) = (a,c) + (b,c)

(a, a) ≥ o)

where equality with zero occurs only for a = o. If vectors a and b have the coordinates $\{X_1, Y_1, Z_1\}$ and $\{X_2, Y_2, Z_2\}$ respectively, in an orthonormalized basis i, j, k, then,

$$(a, b) = X_1 X_2 + Y_1 Y_2 + Z_1 Z_2$$

$$|a| = \sqrt{X_1^2 + Y_1^2 + Z_1^2}$$

$$\cos \varphi = \frac{X_1 X_2 + Y_1 Y_2 + Z_1 Z_2}{\sqrt{X_1^2 + Y_1^2 + Z_1^2} \sqrt{X_2^2 + Y_2^2 + Z_2^2}}$$

The definition of a vector product requires use of the concept of a left- and right-handed ordering of three vectors. The ordered triplet of vectors a, b, c (a is the first vector, b, the second, and c, the third), starting at the same point and not lying in one plane, is called right-handed (left-handed) if the vectors are situated in the same way as the thumb, index, and middle fingers, respectively, of the right (left) hand. Figure shows right-handed (on the right) and left-handed (on the left) triplets of vectors.

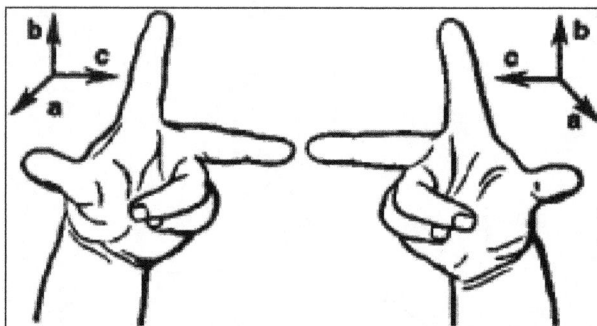

The vector product of vectors a and b is the vector denoted by [a, b] and satisfying the following requirements: (1) the length of vector [a, b] is equal to the product of the lengths of vectors a and and of the sine of the angle φ between them (thus, if a and b are collinear, then [a, b] = o); and (2) if a and b are noncollinear, then [a, b] is perpendicular to both vectors a and b and is directed so that the triplet of vectors a, b, [a, b] is right-handed. The vector product has the following properties:

[a, b] = [b, a]

[(λa), b] = [a, b]

[c, (a + b)] = [c, a] + [c, b]

[a, [b, c]] = b (a, c) - c(a, b)

([a, b], [c, d]) = (a, c)(b, d) - (a, d)(b, c)

If, in an orthonormalized basis i, j, k forming a right-handed triplet, vectors a and b have the coordinates $\{X_1 Y_1 Z_1\}$ and $\{X_2 Y_2 Z_2\}$, respectively, then [a, b] = $\{Y_1 Z_2 - Y_2 Z_1, Z_1 X_2 - Z_2 X_1, X_1 Y_2 - X_2 Y_1\}$. The concept of vector product is connected with various problems in mechanics and physics. For example, the velocity v of a point M of an object rotating around an axis / with an angular velocity of ω is [ω,r], where r = OM.

The mixed product of vectors a, b, and c is the scalar product of vector [a, b] and vector c: ([a, b], c). It is denoted by abc. The mixed product of vectors a, b, and c that are not parallel to the same plane is numerically equal to the volume of the parallelepiped formed by bringing the vectors a, b, and c to a common initial point; its sign is positive if the triplet a, b, c is right-handed and negative if the triplet is left-handed. If vectors a, b, and c are parallel to the same plane, then abc = o. The property that $abc = bca = cab$ also holds true. If the coordinates of vectors a, b, and c in an orthonormalized basis i, j, k, which forms a right-handed triplet, are respectively equal to $\{X_1 Y_1 Z_1\}$, $\{X_2 Y_2 Z_2\}$, and $\{X_3 Y_3 Z_3\}$, then:

$$abc = \begin{vmatrix} X_1 & Y_1 & Z_1 \\ X_2 & Y_2 & Z_2 \\ X_3 & Y_3 & Z_3 \end{vmatrix}$$

Vector Functions of Scalar Arguments

In mechanics, physics, and differential geometry frequent use is made of the concept of a vector function of one or several scalar arguments. If a definite vector r is in correspondence to every value of a variable t of a certain set $\{t\}$ according to a known law, then one says that a vector function r = r(t) is specified by the set $\{t\}$. Since vector r is defined by coordinates $\{x, y, z\}$, the specification of the vector function r = r(t) is equivalent to the specification of three scalar functions: $x = x(t)$, $y = y(t)$, and $z = z(t)$. The concept of vector function becomes particularly obvious if it is converted to a so-called hodograph of this function, that is, to the locus of the ends of all vectors r(t) joined to the coordinate origin O. If, in this case, one considers the argument t to be time, then the vector function r(t) represents the law of motion of point M moving along curve L —the hodograph of r(t).

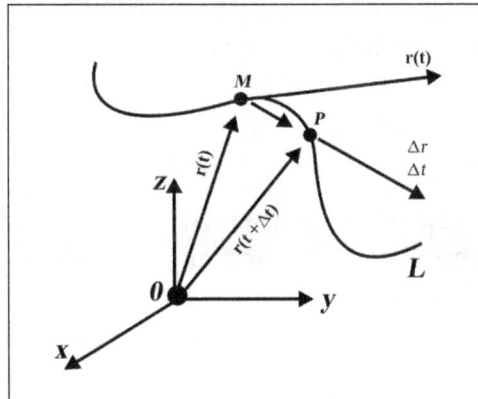

The concept of derivative plays an important role in the study of vector functions. This concept is introduced in the following way: to the argument t is added the increment $\Delta t \neq 0$ and the vector Δr = r(t + Δt) - r(t) (the increment Δr is vector $\overline{MP}$ in figure) is multiplied by $1/\Delta t$. The limit of the expression $\Delta r/\Delta t$ as $\Delta t \to 0$ is called the derivative of the vector function and is denoted by r'(t) or dr/dt. The derivative is the vector that is tangent to the hodograph L at the given point M. If the vector function is regarded as the law of motion of a point along the curve L, then the derivative $r'(t)$ is equal to the velocity of this point's motion. The rules for computing the derivatives of various products of vector functions are similar to the rules of finding the derivatives of the products of ordinary functions.

For example,

$$(r_1, r_2)' = (r_1', r_2) + (r_1, r_2')$$

$$[r_1, r_2]' = [r_1', r_2] + [r_1, r_2']$$

In differential geometry the vector functions of one argument are used for the definition of curves. Vector functions of two arguments are used for the specification of surfaces.

Vector analysis In mechanics, physics, and geometry the concepts of scalar and vector fields are frequently used. The temperature of a nonuniformly heated plate and the density of a nonhomogeneous body are physical examples of plane and three-dimensional scalar fields, respectively. A vector field is a set of all the velocity vectors of particles of a steady flow of fluid. Other examples of vector fields are the gravitational force field and the electrical and magnetic potentials of an electromagnetic field.

For the mathematical specification of scalar and vector fields, scalar and vector functions are used, respectively. It is clear that the density of an object is a scalar point-function and that the velocity field of the particles of steady liquid flow is a vector point-function. The mathematical apparatus of field theory is usually called vector analysis. For the geometric characterization of a scalar field one uses the concepts of contour lines and equipotential surfaces. The contour line of a plane scalar field is a line on which the function that defines the field has a constant value. The equipotential surface of a spatial field is defined in an analogous way. An example of a contour line is an isotherm—the contour line of the scalar temperature field of a nonuniformly heated plate.

We now consider equipotential surfaces (lines) of a scalar field which pass through a given point M. The maximum change of the function f which defines the field at this point, occurs along a normal to this surface (line) at the point M. This change is characterized by the gradient of the scalar field. The gradient is a vector that is directed along the normal to the equipotential surface (line) at point M in the direction of the increasing f at this point. The magnitude of the gradient is equal to the derivative of f in the indicated direction. The gradient is denoted by the symbol grad f. If the basis is i, j, k, then grad f has the coordinates $\left(\partial f / \partial x, \partial f / \partial y, \partial f / \partial z \right)$; for a plane field the gradient coordinates are $\left(\partial f / \partial x, \partial f / \partial y \right)$. The gradient of a scalar field is a vector field.

A number of concepts are introduced to characterize vector fields: vector lines, vector tubes, circulations of a vector field, and divergence and curl (rotor) of a vector field. In some region Ω, let a vector field be denoted by the vector function a*(M)* of a variable point M of Ω. A line L in the region Ω is called a vector line if the vector tangent at each of its points M is directed along vector a*(M)*.

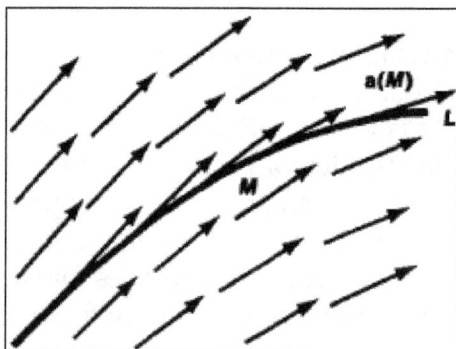

If the field a*(M)* is a velocity field of the particles of a steady flow of a fluid, then the vector lines of this field are the trajectories of the fluid particles. The part of space in Ω that consists of vector lines is called a vector tube. If one is dealing with the vector field of velocities of the particles of a steady flow of a fluid, then the vector tube is the part of space that a certain fixed volume of fluid "sweeps out" in its motion.

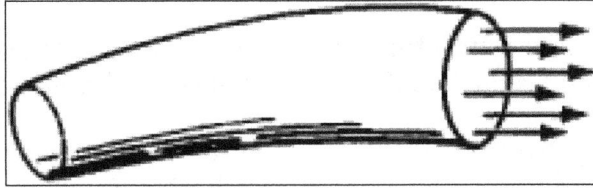

Let *AB* be a smooth curve in Ω, *l* the length of arc *AB* measured from point *A* to the variable point *M* of this line, and t the unit vector tangent to *AB* at *M* . The circulation of the field a*(M)* along the curve *AB* is the expression,

$$\int_{AB} (a, \ )dl$$

If a*(M)* is a force field, then the circulation of a along *AB* is the work performed by this field along the path *AB*.

The divergence of vector field a*(M)* , which has the coordinates *P, Q, R* in the basis i, j, k, is defined as the sum $\partial P/\partial x + \partial Q/\partial y + \partial R/\partial z$ and is denoted by the symbol div a. For example, the divergence of the gravitational field created by a certain mass distribution is equal to the density (volumetric) $\rho(x, y, z)$ of this field multiplied by 4π.

The curl (or rot) of vector field a*(M)* is a vector characterizing the "rotational component" of this field. The curl of field a is denoted by rot a or curl *a* . If *P, Q, R* are the coordinates of a in the basis *i, j, k*, then:

$$\text{rot a} = \left\{ \frac{\partial R}{\partial y} - \frac{\partial Q}{\partial z}, \frac{\partial P}{\partial z} - \frac{\partial R}{\partial x}, \frac{\partial Q}{\partial x} - \frac{\partial P}{\partial y} \right\}$$

Let field a be the velocity field of a fluid flow. We place a small wheel with vanes at a given point of the flow and orient its axis in the direction of rot a at this point. Then the flowrate will be a maximum, and its value will be ½[rot a]. The gradient of a scalar field and the divergence and curl of a vector field are usually called the fundamental differential operations of vector analysis. The following equations, relating these operations, hold true:

grad (*fh*) = *f* grad *h* + *h* grad *f*

div (*f*a) = (a, grad *f*) + *f* div a

rot (*f*a) = *f* rot a + [grad *f* , a]

div [a, b] = (b, rot a) - (a, rot b)

Vector field a*(M)* is called potential if it is the gradient of some scalar field *f(M)* . In this case, the field *f(M)* is called the potential of vector field a. In order that the field a, whose coordinates *P, Q, R* have continuous partial derivatives, be a potential field, it is necessary and sufficient that the curl of this field vanish. If a potential field is given in a simply connected region Ω, then the potential *f(M)* of this field can be found from the formula,

$$f = \int_{AM} (a, t)\, dl$$

Where *AM* is any smooth curve connecting a fixed point *A* of Ω with point *M*, t is the unit vector tangent to the curve *AM,* and *l* is the length of arc *AM* measured from point *A*.

Vector field a*(M)* is called solenoidal or tubular if it is the curl of some field b*(M)*. Field b*(M)* is called the vector potential of field a. In order that a be solenoidal, it is necessary and sufficient that the divergence of this field vanish. An important role in vector analysis is played by integral relations: Ostrogradskii's formula, also designated the fundamental formula of vector analysis, and Stokes' formula. Let *V* be a region whose boundary T consists of a finite number of pieces of smooth surfaces and n be the unit vector of the exterior normal to *T*. Let vector field a*(M)* be given in the region *V* such that div a is a continuous function. Then the following holds true:

$$\iiint_v \operatorname{div} a\, dv = \iint r\, (a, n)\, d\sigma$$

This is known as Ostrogradskii's formula.

If a is the velocity field of a steady flow of incompressible fluid, then (a, n) *dσ* is the volume of fluid that passes through an area *dσ* on the boundary r in a unit of time. Therefore, the right-hand side of equation ($\iiint_v$ div a*dv* = $\iint r$ (a,n) *dσ*) is the flow of fluid through the boundary r of body *V* per unit time. Because, in the case being considered, div a characterizes the intensity of the fluid sources, Ostrogradskii's formula expresses the following obvious fact: the flow of fluid through a closed surface r is equal to the amount of fluid generated by all the sources inside r. Let a continuous and differentiable vector field a which has a continuous curl rot a be assigned in a region Ω. Let r be an orientable surface consisting of a finite number of pieces of smooth surface, n the unit normal to r, t the unit vector tangent to the edge *y* of the surface r, and *l* the length of the arc *y*. The following relation, called Stokes' formula, holds true:

$$\iint_r (n, \operatorname{rot} a)\, d\sigma = \oint_r (a, t)\, dl$$

Equation ($\iint_r (n, \operatorname{rot} a)\, d\sigma = \oint_r (a, t)\, dl$) expresses the following physical fact: the intensity of the curl of a vector field *a* through the surface r is equal to the circulation of this field along the curve *y*. Ostrogradskii's formula is the source of the invariant (independent of the coordinate system) definition of the fundamental operations of vector analysis. For example, from this formula, it follows that:

$$\operatorname{div} a = \lim_{V \to 0} \frac{\iint_r (a, n)\, d\sigma}{V}$$

Because the expression,

$$\iint_r (a, n)\, d\sigma$$

is the flow of fluid through r and

$$\frac{1}{V}\iint (a,n)\,d\sigma$$

is the magnitude of this flow per unit volume, the definition of div a by means of equation $\text{div}\,a = \lim_{V \to 0}\frac{\iint_r (a,n)\,d\sigma}{V}$ indicates that div a characterizes the flux of the source at a given point.

DEL

Del, or nabla, is an operator used in mathematics, in particular in vector calculus, as a vector differential operator, usually represented by the nabla symbol ∇. When applied to a function defined on a one-dimensional domain, it denotes its standard derivative as defined in calculus. When applied to a field (a function defined on a multi-dimensional domain), it may denote the gradient (locally steepest slope) of a scalar field (or sometimes of a vector field, as in the Navier–Stokes equations), the divergence of a vector field, or the curl (rotation) of a vector field, depending on the way it is applied.

Strictly speaking, del is not a specific operator, but rather a convenient mathematical notation for those three operators, that makes many equations easier to write and remember. The del symbol can be interpreted as a vector of partial derivative operators, and its three possible meanings—gradient, divergence, and curl—can be formally viewed as the product with a scalar, a dot product, and a cross product, respectively, of the del "operator" with the field. These formal products do not necessarily commute with other operators or products. These three uses, are summarized as:

- Gradient: $\text{grad}\,f = \nabla f$

- Divergence: $\text{div}\,\vec{v} = \nabla \cdot \vec{v}$

- Curl: $\text{curl}\,\vec{v} = \nabla \times \vec{v}$

In the Cartesian coordinate system R^n with coordinates $(x_1,\dots,x_n)$ and standard basis $\{\vec{e}_1,\dots,\vec{e}_n\}$, del is defined in terms of partial derivative operators as,

$$\nabla = \sum_{i=1}^{n} \vec{e}_i \frac{\partial}{\partial x_i} = \left(\frac{\partial}{\partial x_1},\dots,\frac{\partial}{\partial x_n} \right)$$

In three-dimensional Cartesian coordinate system R^3 with coordinates (x,y,z) and standard basis or unit vectors of axes $\{\vec{e}_x,\vec{e}_y,\vec{e}_z\}$, del is written as,

$$\nabla = \vec{e}_x \frac{\partial}{\partial x} + \vec{e}_y \frac{\partial}{\partial y} + \vec{e}_z \frac{\partial}{\partial z} = \left(\frac{\partial}{\partial x},\frac{\partial}{\partial y},\frac{\partial}{\partial z} \right)$$

Del can also be expressed in other coordinate systems.

Notational Uses

Del is used as a shorthand form to simplify many long mathematical expressions. It is most commonly used to simplify expressions for the gradient, divergence, curl, directional derivative, and Laplacian.

Gradient

The vector derivative of a scalar field f is called the gradient, and it can be represented as:

$$\text{grad } f = \frac{\partial f}{\partial x}\vec{e}_x + \frac{\partial f}{\partial y}\vec{e}_y + \frac{\partial f}{\partial z}\vec{e}_z = \nabla f$$

It always points in the direction of greatest increase of f, and it has a magnitude equal to the maximum rate of increase at the point—just like a standard derivative. In particular, if a hill is defined as a height function over a plane $h(x, y)$, the gradient at a given location will be a vector in the xy-plane (visualizable as an arrow on a map) pointing along the steepest direction. The magnitude of the gradient is the value of this steepest slope.

In particular, this notation is powerful because the gradient product rule looks very similar to the 1d-derivative case:

$$\nabla(fg) = f\nabla g + g\nabla f$$

However, the rules for dot products do not turn out to be simple, as illustrated by:

$$\nabla(\vec{u}\cdot\vec{v}) = (\vec{u}\cdot\nabla)\vec{v} + (\vec{v}\cdot\nabla)\vec{u} + \vec{u}\times(\nabla\times\vec{v}) + \vec{v}\times(\nabla\times\vec{u})$$

Divergence

The divergence of a vector field $\vec{v}(x, y, z) = v_x\vec{e}_x + v_y\vec{e}_y + v_z\vec{e}_z$ is a scalar function that can be represented as:

$$\text{div }\vec{v} = \frac{\partial v_x}{\partial x} + \frac{\partial v_y}{\partial y} + \frac{\partial v_z}{\partial z} = \nabla\cdot\vec{v}$$

The divergence is roughly a measure of a vector field's increase in the direction it points; but more accurately, it is a measure of that field's tendency to converge toward or repel from a point.

The power of the del notation is shown by the following product rule:

$$\nabla\cdot(f\vec{v}) = f(\nabla\cdot\vec{v}) + \vec{v}\cdot(\nabla f)$$

The formula for the vector product is slightly less intuitive, because this product is not commutative:

$$\nabla\cdot(\vec{u}\times\vec{v}) = \vec{v}\cdot(\nabla\times\vec{u}) - \vec{u}\cdot(\nabla\times\vec{v})$$

Curl

The curl of a vector field $\vec{v}(x,y,z) = v_x\vec{e}_x + v_y\vec{e}_y + v_z\vec{e}_z$ is a vector function that can be represented as:

$$\text{curl}\,\vec{v} = \left(\frac{\partial v_z}{\partial y} - \frac{\partial v_y}{\partial z}\right)\vec{e}_x + \left(\frac{\partial v_x}{\partial z} - \frac{\partial v_z}{\partial x}\right)\vec{e}_y + \left(\frac{\partial v_y}{\partial x} - \frac{\partial v_x}{\partial y}\right)\vec{e}_z = \nabla \times \vec{v}$$

The curl at a point is proportional to the on-axis torque that a tiny pinwheel would be subjected to if it were centred at that point.

The vector product operation can be visualized as a pseudo-determinant:

$$\nabla \times \vec{v} = \begin{vmatrix} \vec{e}_x & \vec{e}_y & \vec{e}_z \\ \dfrac{\partial}{\partial x} & \dfrac{\partial}{\partial y} & \dfrac{\partial}{\partial z} \\ v_x & v_y & v_z \end{vmatrix}$$

Again the power of the notation is shown by the product rule:

$$\nabla \times (f\vec{v}) = (\nabla f) \times \vec{v} + f(\nabla \times \vec{v})$$

Unfortunately the rule for the vector product does not turn out to be simple:

$$\nabla \times (\vec{u} \times \vec{v}) = \vec{u}(\nabla \cdot \vec{v}) - \vec{v}(\nabla \cdot \vec{u}) + (\vec{v} \cdot \nabla)\vec{u} - (\vec{u} \cdot \nabla)\vec{v}$$

Directional Derivative

The directional derivative of a scalar field $f(x,y,z)$ in the direction $\vec{a}(x,y,z) = a_x\vec{e}_x + a_y\vec{e}_y + a_z\vec{e}_z$ is defined as:

$$\vec{a} \cdot \text{grad}\,f = a_x\frac{\partial f}{\partial x} + a_y\frac{\partial f}{\partial y} + a_z\frac{\partial f}{\partial z} = \vec{a} \cdot (\nabla f)$$

This gives the rate of change of a field   in the direction of $\vec{a}$. In operator notation, the element in parentheses can be considered a single coherent unit; fluid dynamics uses this convention extensively, terming it the convective derivative—the "moving" derivative of the fluid.

Note that $(\vec{a} \cdot \nabla)$ is an operator that takes scalar to a scalar. It can be extended to operate on a vector, by separately operating on each of its components.

Laplacian

The Laplace operator is a scalar operator that can be applied to either vector or scalar fields; for cartesian coordinate systems it is defined as:

$$\Delta = \frac{\partial^2}{\partial x^2} + \frac{\partial^2}{\partial y^2} + \frac{\partial^2}{\partial z^2} = \nabla \cdot \nabla = \nabla^2$$

and the definition for more general coordinate systems is given in vector Laplacian.

The Laplacian is ubiquitous throughout modern mathematical physics, appearing for example in Laplace's equation, Poisson's equation, the heat equation, the wave equation, and the Schrödinger equation.

Hessian Matrix

While ∇^2 usually represents the Laplacian, sometimes ∇^2 also represents the Hessian matrix. The former refers to the inner product of ∇, while the latter refers to the dyadic product of ∇:

$$\nabla^2 = \nabla \cdot \nabla^T.$$

So whether ∇^2 refers to a Laplacian or a Hessian matrix depends on the context.

Tensor Derivative

Del can also be applied to a vector field with the result being a tensor. The tensor derivative of a vector field $\vec{v}$ (in three dimensions) is a 9-term second-rank tensor – that is, a 3×3 matrix – but can be denoted simply as $\nabla \otimes \vec{v}$, where $\otimes$ represents the dyadic product. This quantity is equivalent to the transpose of the Jacobian matrix of the vector field with respect to space. The divergence of the vector field can then be expressed as the trace of this matrix.

For a small displacement $\delta\vec{r}$, the change in the vector field is given by:

$$\delta\vec{v} = (\nabla \otimes \vec{v})^T \cdot \delta\vec{r}$$

Product Rules

For vector calculus:

$$\nabla(fg) = f\nabla g + g\nabla f$$
$$\nabla(\vec{u} \cdot \vec{v}) = \vec{u} \times (\nabla \times \vec{v}) + \vec{v} \times (\nabla \times \vec{u}) + (\vec{u} \cdot \nabla)\vec{v} + (\vec{v} \cdot \nabla)\vec{u}$$
$$\nabla \cdot (f\vec{v}) = f(\nabla \cdot \vec{v}) + \vec{v} \cdot (\nabla f)$$
$$\nabla \cdot (\vec{u} \times \vec{v}) = \vec{v} \cdot (\nabla \times \vec{u}) - \vec{u} \cdot (\nabla \times \vec{v})$$
$$\nabla \times (f\vec{v}) = (\nabla f) \times \vec{v} + f(\nabla \times \vec{v})$$
$$\nabla \times (\vec{u} \times \vec{v}) = \vec{u}(\nabla \cdot \vec{v}) - \vec{v}(\nabla \cdot \vec{u}) + (\vec{v} \cdot \nabla)\vec{u} - (\vec{u} \cdot \nabla)\vec{v}$$

For matrix calculus (for which $\vec{u} \cdot \vec{v}$ can be written $\vec{u}^T\vec{v}$):

$$(A\nabla)^T\vec{u} = \nabla^T(A^T\vec{u}) - (\nabla^T A^T)\vec{u}$$

Another relation of interest is the following, where $\vec{u} \otimes \vec{v}$ is the outer product tensor:

$$\nabla \cdot (\vec{u} \otimes \vec{v}) = (\nabla \cdot \vec{u})\vec{v} + (\vec{u} \cdot \nabla)\vec{v}$$

Second Derivatives

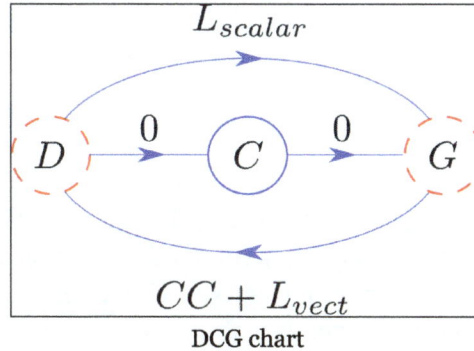

DCG chart

In figure, a simple chart depicting all rules pertaining to second derivatives. D, C, G, L and CC stand for divergence, curl, gradient, Laplacian and curl of curl, respectively. Arrows indicate existence of second derivatives. Blue circle in the middle represents curl of curl, whereas the other two red circles (dashed) mean that DD and GG do not exist.

When del operates on a scalar or vector, either a scalar or vector is returned. Because of the diversity of vector products (scalar, dot, cross) one application of del already gives rise to three major derivatives: the gradient (scalar product), divergence (dot product), and curl (cross product). Applying these three sorts of derivatives again to each other gives five possible second derivatives, for a scalar field f or a vector field v; the use of the scalar Laplacian and vector Laplacian gives two more:

$$\mathrm{div}(\mathrm{grad}\, f) = \nabla \cdot (\nabla f)$$
$$\mathrm{curl}(\mathrm{grad}\, f) = \nabla \times (\nabla f)$$
$$\Delta f = \nabla^2 f$$
$$\mathrm{grad}(\mathrm{div}\, \vec{v}) = \nabla(\nabla \cdot \vec{v})$$
$$\mathrm{div}(\mathrm{curl}\, \vec{v}) = \nabla \cdot (\nabla \times \vec{v})$$
$$\mathrm{curl}(\mathrm{curl}\, \vec{v}) = \nabla \times (\nabla \times \vec{v})$$
$$\Delta \vec{v} = \nabla^2 \vec{v}$$

These are of interest principally because they are not always unique or independent of each other. As long as the functions are well-behaved, two of them are always zero:

$$\mathrm{curl}(\mathrm{grad}\, f) = \nabla \times (\nabla f) = 0$$
$$\mathrm{div}(\mathrm{curl}\, \vec{v}) = \nabla \cdot \nabla \times \vec{v} = 0$$

Two of them are always equal:

$$\mathrm{div}(\mathrm{grad}\, f) = \nabla \cdot (\nabla f) = \nabla^2 f = \Delta f$$

The 3 remaining vector derivatives are related by the equation:

$$\nabla \times (\nabla \times \vec{v}) = \nabla(\nabla \cdot \vec{v}) - \nabla^2 \vec{v}$$

And one of them can even be expressed with the tensor product, if the functions are well-behaved:

$$\nabla(\nabla\cdot\vec{v}) = \nabla\cdot(\nabla\otimes\vec{v})$$

Precautions

Most of the above vector properties (except for those that rely explicitly on del's differential properties—for example, the product rule) rely only on symbol rearrangement, and must necessarily hold if the del symbol is replaced by any other vector. This is part of the value to be gained in notationally representing this operator as a vector.

Though one can often replace del with a vector and obtain a vector identity, making those identities mnemonic, the reverse is *not* necessarily reliable, because del does not commute in general.

A counterexample that relies on del's failure to commute:

$$(\vec{u}\cdot\vec{v})f \equiv (\vec{v}\cdot\vec{u})f$$

$$(\nabla\cdot\vec{v})f = \left(\frac{\partial v_x}{\partial x} + \frac{\partial v_y}{\partial y} + \frac{\partial v_z}{\partial z}\right)f = \frac{\partial v_x}{\partial x}f + \frac{\partial v_y}{\partial y}f + \frac{\partial v_z}{\partial z}f$$

$$(\vec{v}\cdot\nabla)f = \left(v_x\frac{\partial}{\partial x} + v_y\frac{\partial}{\partial y} + v_z\frac{\partial}{\partial z}\right)f = v_x\frac{\partial f}{\partial x} + v_y\frac{\partial f}{\partial y} + v_z\frac{\partial f}{\partial z}$$

$$\Rightarrow (\nabla\cdot\vec{v})f \neq (\vec{v}\cdot\nabla)f$$

A counterexample that relies on del's differential properties:

$$(\nabla x)\times(\nabla y) = \left(\vec{e}_x\frac{\partial x}{\partial x} + \vec{e}_y\frac{\partial x}{\partial y} + \vec{e}_z\frac{\partial x}{\partial z}\right)\times\left(\vec{e}_x\frac{\partial y}{\partial x} + \vec{e}_y\frac{\partial y}{\partial y} + \vec{e}_z\frac{\partial y}{\partial z}\right)$$

$$= (\vec{e}_x\cdot 1 + \vec{e}_y\cdot 0 + \vec{e}_z\cdot 0)\times(\vec{e}_x\cdot 0 + \vec{e}_y\cdot 1 + \vec{e}_z\cdot 0)$$

$$= \vec{e}_x\times\vec{e}_y$$

$$= \vec{e}_z$$

$$(\vec{u}x)\times(\vec{u}y) = xy(\vec{u}\times\vec{u})$$

$$= xy\vec{0}$$

$$= \vec{0}$$

Central to these distinctions is the fact that del is not simply a vector; it is a vector operator. Whereas a vector is an object with both a magnitude and direction, del has neither a magnitude nor a direction until it operates on a function.

For that reason, identities involving del must be derived with care, using both vector identities and *differentiation* identities such as the product rule.

GRADIENT

In vector calculus, the gradient is a multi-variable generalization of the derivative. Whereas the ordinary derivative of a function of a single variable is a scalar-valued function, the gradient of a function of several variables is a vector-valued function. Specifically, the gradient of a differentiable function f of several variables, at a point P, is the vector whose components are the partial derivatives of f at P.

Much as the derivative of a function of a single variable represents the slope of the tangent to the graph of the function, if at a point P, the gradient of a function of several variables is not the zero vector, it has the direction of greatest increase of the function at P, and its magnitude is the rate of increase in that direction.

The magnitude and direction of the gradient vector are independent of the particular coordinate representation.

The Jacobian is the generalization of the gradient for vector-valued functions of several variables and differentiable maps between Euclidean spaces or, more generally, manifolds. A further generalization for a function between Banach spaces is the Fréchet derivative.

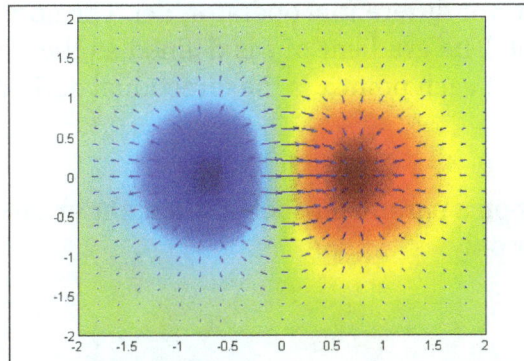

Gradient of the 2D function $f(x, y) = xe^{-(x^2 + y^2)}$ is plotted as blue arrows over the pseudocolor plot of the function.

Consider a room where the temperature is given by a scalar field, T, so at each point (x, y, z) the temperature is $T(x, y, z)$. (Assume that the temperature does not change over time.) At each point in the room, the gradient of T at that point will show the direction in which the temperature rises most quickly. The magnitude of the gradient will determine how fast the temperature rises in that direction.

Consider a surface whose height above sea level at point (x, y) is $H(x, y)$. The gradient of H at a point is a vector pointing in the direction of the steepest slope or grade at that point. The steepness of the slope at that point is given by the magnitude of the gradient vector.

The gradient can also be used to measure how a scalar field changes in other directions, rather than just the direction of greatest change, by taking a dot product. Suppose that the steepest slope on a hill is 40%. If a road goes directly up the hill, then the steepest slope on the road will also be 40%. If, instead, the road goes around the hill at an angle, then it will have a shallower slope. For example, if the angle between the road and the uphill direction, projected onto the horizontal plane, is 60°, then the steepest slope along the road will be 20%, which is 40% times the cosine of 60°.

This observation can be mathematically stated as follows. If the hill height function H is differentiable, then the gradient of H dotted with a unit vector gives the slope of the hill in the direction of the vector. More precisely, when H is differentiable, the dot product of the gradient of H with a given unit vector is equal to the directional derivative of H in the direction of that unit vector.

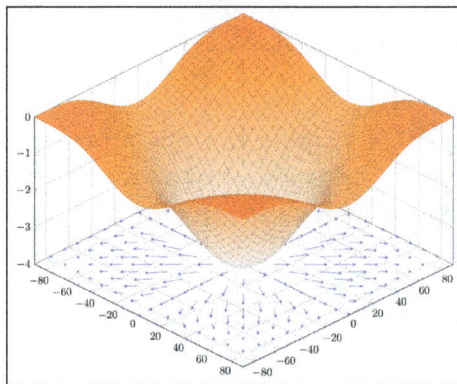

The gradient of the function $f(x,y) = -(\cos^2 x + \cos^2 y)^2$ depicted as a projected vector field on the bottom plane.

The gradient (or gradient vector field) of a scalar function $f(x_1, x_2, x_3, ..., x_n)$ is denoted ∇f or $\vec{\nabla} f$ where ∇ (nabla) denotes the vector differential operator, del. The notation $\mathrm{grad}\, f$ is also commonly used to represent the gradient. The gradient of f is defined as the unique vector field whose dot product with any unit vector v at each point x is the directional derivative of f along v. That is,

$$\left(\nabla f(x)\right)\cdot \mathrm{v} = D_{\mathrm{v}} f(x).$$

When a function also depends on a parameter such as time, the gradient often refers simply to the vector of its spatial derivatives only.

Cartesian Coordinates

In the three-dimensional Cartesian coordinate system with a Euclidean metric, the gradient, if it exists, is given by:

$$\nabla f = \frac{\partial f}{\partial x}\mathrm{i} + \frac{\partial f}{\partial y}\mathrm{j} + \frac{\partial f}{\partial z}\mathrm{k},$$

where i, j, k are the standard unit vectors in the directions of the x, y and z coordinates, respectively. For example, the gradient of the function,

$$f(x, y, z) = 2x + 3y^2 - \sin(z)$$

is,

$$\nabla f = 2\mathrm{i} + 6y\mathrm{j} - \cos(z)\,\mathrm{k}.$$

In some applications it is customary to represent the gradient as a row vector or column vector of its components in a rectangular coordinate system.

Cylindrical and Spherical Coordinates

In cylindrical coordinates with a Euclidean metric, the gradient is given by:

$$\nabla f(\rho, \varphi, z) = \frac{\partial f}{\partial \rho} e_\rho + \frac{1}{\rho} \frac{\partial f}{\partial \varphi} e_\varphi + \frac{\partial f}{\partial z} e_z,$$

where ρ is the axial distance, φ is the azimuthal or azimuth angle, z is the axial coordinate, and e_ρ, e_φ and e_z are unit vectors pointing along the coordinate directions.

In spherical coordinates, the gradient is given by:

$$\nabla f(r, \theta, \varphi) = \frac{\partial f}{\partial r} e_r + \frac{1}{r} \frac{\partial f}{\partial \theta} e_\theta + \frac{1}{r \sin \theta} \frac{\partial f}{\partial \varphi} e_\varphi,$$

where r is the radial distance, φ is the azimuthal angle and θ is the polar angle, and e_r, e_θ and e_φ are again local unit vectors pointing in the coordinate directions (i.e. the normalized covariant basis).

General Coordinates

We consider general coordinates, which we write as x^1, ..., x^i, ..., x^n, where n is the number of dimensions of the domain. Here, the upper index refers to the position in the list of the coordinate or component, so x^2 refers to the second component—not the quantity x squared. The index variable i refers to an arbitrary element x^i. Using Einstein notation, the gradient can then be written as:

$$\nabla f = \frac{\partial f}{\partial x^i} g^{ij} e_j \quad (\text{ Note that its dual is } df = \frac{\partial f}{\partial x^i} e^i),$$

where $e_i = \partial x / \partial x^i$ and $e^i = dx^i$ refer to the unnormalized local covariant and contravariant bases respectively, g^{ij} is the inverse metric tensor, and the Einstein summation convention implies summation over i and j.

If the coordinates are orthogonal we can easily express the gradient (and the differential) in terms of the normalized bases, which we refer to as $\hat{e}_i$ and $\hat{e}^i$, using the scale factors (also known as Lamé coefficients) $h_i = \| e_i \| = 1 / \| e^i \|$:

$$\nabla f = \sum_{i=1}^{n} \frac{\partial f}{\partial x^i} \frac{1}{h_i} \hat{e}_i \quad (\text{ and } df = \sum_{i=1}^{n} \frac{\partial f}{\partial x^i} \frac{1}{h_i} \hat{e}^i),$$

where we cannot use Einstein notation, since it is impossible to avoid the repetition of more than two indices. Despite the use of upper and lower indices, $\hat{e}_i$, $\hat{e}^i$, and h_i are neither contravariant nor covariant.

The latter expression evaluates to the expressions given above for cylindrical and spherical coordinates.

Gradient and the Derivative or Differential

Linear Approximation to a Function

The gradient of a function f from the Euclidean space R^n to R at any particular point x_0 in R^n characterizes the best linear approximation to f at x_0. The approximation is as follows:

$$f(x) \approx f(x_0) + (\nabla f)_{x_0} \cdot (x - x_0)$$

for x close to x_0, where $(\nabla f)_{x0}$ is the gradient of f computed at x_0, and the dot denotes the dot product on R^n. This equation is equivalent to the first two terms in the multivariable Taylor series expansion of f at x_0.

Differential or (Exterior) Derivative

The best linear approximation to a differentiable function,

$$f : \mathrm{R}^n \to \mathrm{R}$$

at a point x in R^n is a linear map from R^n to R which is often denoted by df_x or $Df(x)$ and called the differential or (total) derivative of f at x. The gradient is therefore related to the differential by the formula,

$$(\nabla f)_x \cdot v = df_x(v)$$

for any $v \in \mathrm{R}^n$. The function df, which maps x to df_x, is called the differential or exterior derivative of f and is an example of a differential 1-form.

If R^n is viewed as the space of (dimension n) column vectors (of real numbers), then one can regard df as the row vector with components,

$$\left(\frac{\partial f}{\partial x_1}, \ldots, \frac{\partial f}{\partial x_n} \right),$$

so that $df_x(v)$ is given by matrix multiplication. Assuming the standard Euclidean metric on R^n, the gradient is then the corresponding column vector, i.e.,

$$(\nabla f)_i = df_i^{\mathsf{T}}.$$

Gradient as a Derivative

Let U be an open set in R^n. If the function $f : U \to \mathrm{R}$ is differentiable, then the differential of f is the (Fréchet) derivative of f. Thus ∇f is a function from U to the space R^n such that,

$$\lim_{h \to 0} \frac{|f(x+h) - f(x) - \nabla f(x) \cdot h|}{\|h\|} = 0,$$

where · is the dot product.

As a consequence, the usual properties of the derivative hold for the gradient:

Linearity

The gradient is linear in the sense that if f and g are two real-valued functions differentiable at the point $a \in R^n$, and α and β are two constants, then $\alpha f + \beta g$ is differentiable at a, and moreover,

$$\nabla(\alpha f + \beta g)(a) = \alpha \nabla f(a) + \beta \nabla g(a).$$

Product Rule

If f and g are real-valued functions differentiable at a point $a \in R^n$, then the product rule asserts that the product fg is differentiable at a, and

$$\nabla(fg)(a) = f(a)\nabla g(a) + g(a)\nabla f(a).$$

Chain Rule

Suppose that $f : A \rightarrow R$ is a real-valued function defined on a subset A of R^n, and that f is differentiable at a point a. There are two forms of the chain rule applying to the gradient. First, suppose that the function g is a parametric curve; that is, a function $g : I \rightarrow R^n$ maps a subset $I \subset R$ into R^n. If g is differentiable at a point $c \in I$ such that $g(c) = a$, then:

$$(f \circ g)'(c) = \nabla f(a) \cdot g'(c),$$

where $\circ$ is the composition operator: $(f \circ g)(x) = f(g(x))$.

More generally, if instead $I \subset R^k$, then the following holds:

$$\nabla(f \circ g)(c) = \big(Dg(c)\big)^T \big(\nabla f(a)\big),$$

where $(Dg)^T$ denotes the transpose Jacobian matrix.

For the second form of the chain rule, suppose that $h : I \rightarrow R$ is a real valued function on a subset I of R, and that h is differentiable at the point $f(a) \in I$. Then:

$$\nabla(h \circ f)(a) = h'\big(f(a)\big)\nabla f(a).$$

Further Properties and Applications

Level Sets

A level surface, or isosurface, is the set of all points where some function has a given value.

If f is differentiable, then the dot product $(\nabla f)_x \cdot v$ of the gradient at a point x with a vector v gives the directional derivative of f at x in the direction v. It follows that in this case the gradient of f is orthogonal to the level sets of f. For example, a level surface in three-dimensional space is defined by an equation of the form $F(x, y, z) = c$. The gradient of F is then normal to the surface.

More generally, any embedded hypersurface in a Riemannian manifold can be cut out by an equation of the form $F(P) = 0$ such that dF is nowhere zero. The gradient of F is then normal to the hypersurface.

Similarly, an affine algebraic hypersurface may be defined by an equation $F(x_1, ..., x_n) = 0$, where F is a polynomial. The gradient of F is zero at a singular point of the hypersurface (this is the definition of a singular point). At a non-singular point, it is a nonzero normal vector.

Conservative Vector Fields and the Gradient Theorem

The gradient of a function is called a gradient field. A (continuous) gradient field is always a conservative vector field: its line integral along any path depends only on the endpoints of the path, and can be evaluated by the gradient theorem (the fundamental theorem of calculus for line integrals). Conversely, a (continuous) conservative vector field is always the gradient of a function.

Generalizations

Gradient of a Vector

Since the total derivative of a vector field is a linear mapping from vectors to vectors, it is a tensor quantity.

In rectangular coordinates, the gradient of a vector field $\mathbf{f} = (f^1, f^2, f^3)$ is defined by:

$$\nabla \mathbf{f} = g^{jk} \frac{\partial f^i}{\partial x^j} \mathbf{e}_i \otimes \mathbf{e}_k,$$

(where the Einstein summation notation is used and the tensor product of the vectors $\mathbf{e}_i$ and $\mathbf{e}_k$ is a dyadic tensor of type (2,0)). Overall, this expression equals the transpose of the Jacobian matrix:

$$\frac{\partial f^i}{\partial x^j} = \frac{\partial(f^1, f^2, f^3)}{\partial(x^1, x^2, x^3)}.$$

In curvilinear coordinates, or more generally on a curved manifold, the gradient involves Christoffel symbols:

$$\nabla \mathbf{f} = g^{jk}\left(\frac{\partial f^i}{\partial x^j} + \Gamma^i_{jl} f^l\right) \mathbf{e}_i \otimes \mathbf{e}_k,$$

where g^{jk} are the components of the inverse metric tensor and the $\mathbf{e}_i$ are the coordinate basis vectors.

Expressed more invariantly, the gradient of a vector field f can be defined by the Levi-Civita connection and metric tensor:

$$\nabla^a f^b = g^{ac} \nabla_c f^b,$$

where ∇_c is the connection.

Riemannian Manifolds

For any smooth function f on a Riemannian manifold (M, g), the gradient of f is the vector field ∇f such that for any vector field X,

$$g(\nabla f, X) = \partial_X f,$$

i.e.,

$$g_x\big((\nabla f)_x, X_x\big) = (\partial_X f)(x),$$

where $g_x(\ ,\ )$ denotes the inner product of tangent vectors at x defined by the metric g and $\partial_X f$ is the function that takes any point $x \in M$ to the directional derivative of f in the direction X, evaluated at x. In other words, in a coordinate chart φ from an open subset of M to an open subset of R^n, $(\partial_X f)(x)$ is given by:

$$\sum_{j=1}^{n} X^j\big(\varphi(x)\big) \frac{\partial}{\partial x_j} (f \circ \varphi^{-1})\big|_{\varphi(x)},$$

where X^j denotes the jth component of X in this coordinate chart.

So, the local form of the gradient takes the form:

$$\nabla f = g^{ik} \frac{\partial f}{\partial x^k} \mathrm{e}_i.$$

Generalizing the case $M = \mathrm{R}^n$, the gradient of a function is related to its exterior derivative, since

$$(\partial_X f)(x) = (df)_x(X_x).$$

More precisely, the gradient ∇f is the vector field associated to the differential 1-form df using the musical isomorphism

$$\sharp = \sharp^g : T^*M \to TM$$

(called "sharp") defined by the metric g. The relation between the exterior derivative and the gradient of a function on R^n is a special case of this in which the metric is the flat metric given by the dot product.

DIVERGENCE

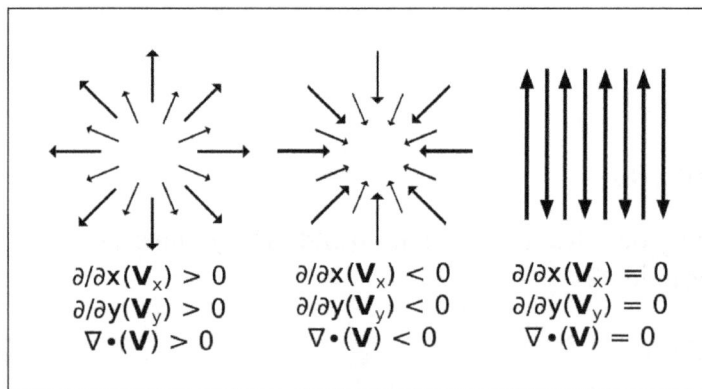

$$\partial/\partial x(\mathbf{V}_x) > 0 \qquad \partial/\partial x(\mathbf{V}_x) < 0 \qquad \partial/\partial x(\mathbf{V}_x) = 0$$
$$\partial/\partial y(\mathbf{V}_y) > 0 \qquad \partial/\partial y(\mathbf{V}_y) < 0 \qquad \partial/\partial y(\mathbf{V}_y) = 0$$
$$\nabla\cdot(\mathbf{V}) > 0 \qquad \nabla\cdot(\mathbf{V}) < 0 \qquad \nabla\cdot(\mathbf{V}) = 0$$

In figure, the divergence of different vector fields. The divergence of vectors from point (x,y) equals the sum of the partial derivative-with-respect-to-x of the x-component and the partial derivative-with-respect-to-y of the y-component at that point:

$$\nabla\cdot(V(x,y)) = \frac{\partial\, V_x(x,y)}{\partial x} + \frac{\partial\, V_y(x,y)}{\partial y}$$

In vector calculus, divergence is a vector operator that produces a scalar field, giving the quantity of a vector field's source at each point. More technically, the divergence represents the volume density of the outward flux of a vector field from an infinitesimal volume around a given point.

As an example, consider air as it is heated or cooled. The velocity of the air at each point defines a vector field. While air is heated in a region, it expands in all directions, and thus the velocity field points outward from that region. The divergence of the velocity field in that region would thus have a positive value. While the air is cooled and thus contracting, the divergence of the velocity has a negative value.

Physical Interpretation of Divergence

In physical terms, the divergence of a three-dimensional vector field is the extent to which the vector field flux behaves like a source at a given point. It is a local measure of its "outgoingness" – the extent to which there is more of some quantity exiting an infinitesimal region of space than entering it. If the divergence is nonzero at some point then there is compression or expansion at that point.

More rigorously, the divergence of a vector field F at a point p can be defined as the limit of the net flux of F across the smooth boundary of a three-dimensional region V divided by the volume of V as V shrinks to p. Formally,

$$\operatorname{div}\mathrm{F}\big|_p = \lim_{V\to\{p\}} \iint_{S(V)} \frac{\mathrm{F}\cdot\hat{\mathrm{n}}}{|V|} dS,$$

where $|V|$ is the volume of V, $S(V)$ is the boundary of V, and the integral is a surface integral with n̂ being the outward unit normal to that surface. The result, div F, is a function of p. From this definition it also becomes obvious that div F can be seen as the *source density* of the flux of F.

In light of the physical interpretation, a vector field with zero divergence everywhere is called *incompressible* or *solenoidal* – in which case any closed surface has no net flux across it.

The intuition that the sum of all sources minus the sum of all sinks should give the net flux outwards of a region is made precise by the divergence theorem.

Cartesian Coordinates

In three-dimensional Cartesian coordinates, the divergence of a continuously differentiable vector field $F = F_x \mathbf{i} + F_y \mathbf{j} + F_z \mathbf{k}$ is defined as the scalar-valued function:

$$\operatorname{div} F = \nabla \cdot F = \left(\frac{\partial}{\partial x}, \frac{\partial}{\partial y}, \frac{\partial}{\partial z} \right) \cdot (F_x, F_y, F_z) = \frac{\partial F_x}{\partial x} + \frac{\partial F_y}{\partial y} + \frac{\partial F_z}{\partial z}.$$

Although expressed in terms of coordinates, the result is invariant under rotations, as the physical interpretation suggests. This is because the trace of the Jacobian matrix of an N-dimensional vector field F in N-dimensional space is invariant under any invertible linear transformation.

The common notation for the divergence $\nabla \cdot F$ is a convenient mnemonic, where the dot denotes an operation reminiscent of the dot product: take the components of the ∇ operator, apply them to the corresponding components of F, and sum the results. Because applying an operator is different from multiplying the components, this is considered an abuse of notation.

The divergence of a continuously differentiable second-order tensor field ε is a first-order tensor field:

$$\overrightarrow{\operatorname{div}}(\mathring{a}) = \begin{bmatrix} \dfrac{\partial \varepsilon_{xx}}{\partial x} + \dfrac{\partial \varepsilon_{yx}}{\partial y} + \dfrac{\partial \varepsilon_{zx}}{\partial z} \\[2ex] \dfrac{\partial \varepsilon_{xy}}{\partial x} + \dfrac{\partial \varepsilon_{yy}}{\partial y} + \dfrac{\partial \varepsilon_{zy}}{\partial z} \\[2ex] \dfrac{\partial \varepsilon_{xz}}{\partial x} + \dfrac{\partial \varepsilon_{yz}}{\partial y} + \dfrac{\partial \varepsilon_{zz}}{\partial z} \end{bmatrix}.$$

Cylindrical Coordinates

For a vector expressed in local unit cylindrical coordinates as,

$$F = e_r F_r + e_\theta F_\theta + e_z F_z,$$

where e_a is the unit vector in direction a, the divergence is,

$$\operatorname{div} F = \nabla \cdot F = \frac{1}{r} \frac{\partial}{\partial r} (r F_r) + \frac{1}{r} \frac{\partial F_\theta}{\partial \theta} + \frac{\partial F_z}{\partial z}.$$

The use of local coordinates is vital for the validity of the expression. If we consider x the position vector and the functions $r(\mathrm{x})$, $\theta(\mathrm{x})$, and $z(\mathrm{x})$, which assign the corresponding global cylindrical coordinate to a vector, in general $r(\mathrm{F(x)}) \neq F_r(\mathrm{x})$, $\theta(\mathrm{F(x)}) \neq F_\theta(\mathrm{x})$, and $z(\mathrm{F(x)}) \neq F_z(\mathrm{x})$. In particular, if we consider the identity function $\mathrm{F(x)} = \mathrm{x}$, we find that:

$$\theta(\mathrm{F(x)}) = \theta \neq F_\theta(\mathrm{x}) = 0.$$

Spherical Coordinates

In spherical coordinates, with θ the angle with the z axis and φ the rotation around the z axis, and F again written in local unit coordinates, the divergence is,

$$\mathrm{div\,F} = \nabla \cdot \mathrm{F} = \frac{1}{r^2} \frac{\partial}{\partial r}\left(r^2 F_r\right) + \frac{1}{r \sin\theta} \frac{\partial}{\partial\theta}(\sin\theta F_\theta) + \frac{1}{r \sin\theta} \frac{\partial F_\varphi}{\partial\varphi}.$$

General Coordinates

Using Einstein notation we can consider the divergence in general coordinates, which we write as x^1, ..., x^i, ...,x^n, where n is the number of dimensions of the domain. Here, the upper index refers to the number of the coordinate or component, so x^2 refers to the second component, and not the quantity x squared. The index variable i is used to refer to an arbitrary element, such as x^i. The divergence can then be written via the Voss- Weyl formula, as:

$$\mathrm{div(F)} = \frac{1}{\rho} \frac{\partial\left(\rho F^i\right)}{\partial x^i}$$

where ρ is the local coefficient of the volume element and F^i are the components of F with respect to the local unnormalized covariant basis (sometimes written as $\mathrm{e}_i = \partial \mathrm{x} / \partial x^i$). The Einstein notation implies summation over i, since it appears as both an upper and lower index.

The volume coefficient ρ is a function of position which depends on the coordinate system. In Cartesian, cylindrical and spherical coordinates, using the same conventions as before, we have $\rho = 1$, $\rho = r$ and $\rho = r^2 \sin\theta$, respectively. It can also be expressed as $\rho = \sqrt{\det g_{ab}}$, where g_{ab} is the metric tensor. Since the determinant is a scalar quantity which doesn't depend on the indices, we can suppress them and simply write $\rho = \sqrt{\det g}$. Another expression comes from computing the determinant of the Jacobian for transforming from Cartesian coordinates, which for $n = 3$ gives,

$$\rho = \left| \frac{\partial(x, y, z)}{\partial(x^1, x^2, x^3)} \right|.$$

Some conventions expect all local basis elements to be normalized to unit length. If we write $\hat{e}_i$ for the normalized basis, and $\hat{F}^i$ for the components of F with respect to it, we have that,

$$\mathrm{F} = F^i \mathrm{e}_i = F^i \|\mathrm{e}_i\| \frac{\mathrm{e}_i}{\|\mathrm{e}_i\|} = F^i \sqrt{g_{ii}}\, \hat{e}_i = \hat{F}^i \hat{e}_i,$$

using one of the properties of the metric tensor. By dotting both sides of the last equality with the contravariant element $\hat{e}_i$, we can conclude that $F^i = \hat{F}^i / \sqrt{g_{ii}}$. After substituting, the formula becomes:

$$\text{div}(F) = \frac{1}{\rho} \frac{\partial \left(\frac{\rho}{\sqrt{g_{ii}}} \hat{F}^i \right)}{\partial x^i} = \frac{1}{\sqrt{\det g}} \frac{\partial \left(\sqrt{\frac{\det g}{g_{ii}}} \hat{F}^i \right)}{\partial x^i}.$$

Decomposition Theorem

It can be shown that any stationary flux v(r) that is at least twice continuously differentiable in R³ and vanishes sufficiently fast for $|r| \to \infty$ can be decomposed into an *irrotational part* E(r) and a *source-free part* B(r). Moreover, these parts are explicitly determined by the respective source densities and circulation densities:

For the irrotational part one has,

$$E = -\nabla \Phi(r),$$

with,

$$\Phi(r) = \int_{\mathbb{R}^3} d^3 r' \frac{\text{div } v(r')}{4\pi |r - r'|}.$$

The source-free part, B, can be similarly written: one only has to replace the *scalar potential* $\Phi(r)$ by a *vector potential* A(r) and the terms $-\nabla \Phi$ by $+\nabla \times A$, and the source density div v by the circulation density $\nabla \times v$.

This "decomposition theorem" is a by-product of the stationary case of electrodynamics. It is a special case of the more general Helmholtz decomposition which works in dimensions greater than three as well.

Properties

The following properties can all be derived from the ordinary differentiation rules of calculus. Most importantly, the divergence is a linear operator, i.e.,

$$\text{div}(a\,F + b\,G) = a\,\text{div}\,F + b\,\text{div}\,G$$

for all vector fields F and G and all real numbers a and b.

There is a product rule of the following type: if φ is a scalar-valued function and F is a vector field, then,

$$\text{div}(\varphi\,F) = \text{grad}\,\varphi \cdot F + \varphi\,\text{div}\,F,$$

or in more suggestive notation,

$$\nabla \cdot (\varphi\,F) = (\nabla \varphi) \cdot F + \varphi(\nabla \cdot F).$$

Another product rule for the cross product of two vector fields F and G in three dimensions involves the curl and reads as follows:

$$\text{div}(F \times G) = \text{curl}\, F \cdot G - F \cdot \text{curl}\, G,$$

or

$$\nabla \cdot (F \times G) = (\nabla \times F) \cdot G - F \cdot (\nabla \times G).$$

The Laplacian of a scalar field is the divergence of the field's gradient:

$$\text{div}(\nabla \varphi) = \Delta \varphi.$$

The divergence of the curl of any vector field (in three dimensions) is equal to zero:

$$\nabla \cdot (\nabla \times F) = 0.$$

If a vector field F with zero divergence is defined on a ball in R³, then there exists some vector field G on the ball with F = curl G. For regions in R³ more topologically complicated than this, the latter statement might be false. The degree of *failure* of the truth of the statement, measured by the homology of the chain complex,

$$\{\text{scalar fields on } U\} \xrightarrow{\text{grad}} \{\text{vector fields on } U\} \xrightarrow{\text{curl}} \{\text{vector fields on } U\} \xrightarrow{\text{div}} \{\text{scalar fields on } U\}$$

serves as a nice quantification of the complicatedness of the underlying region U. These are the beginnings and main motivations of de Rham cohomology.

Relation with the Exterior Derivative

One can express the divergence as a particular case of the exterior derivative, which takes a 2-form to a 3-form in R³. Define the current two-form as,

$$j = F_1\, dy \wedge dz + F_2\, dz \wedge dx + F_3\, dx \wedge dy.$$

It measures the amount of "stuff" flowing through a surface per unit time in a "stuff fluid" of density $\rho = 1\, dx \wedge dy \wedge dz$ moving with local velocity F. Its exterior derivative dj is then given by,

$$dj = \left(\frac{\partial F_1}{\partial x} + \frac{\partial F_2}{\partial y} + \frac{\partial F_3}{\partial z} \right) dx \wedge dy \wedge dz = (\nabla \cdot F)\rho.$$

Thus, the divergence of the vector field F can be expressed as:

$$\nabla \cdot F = \star d \star \left(F^{\flat} \right).$$

Here the superscript $\flat$ is one of the two musical isomorphisms, and $\star$ is the Hodge star operator.

Working with the current two-form and the exterior derivative is usually easier than working with the vector field and divergence, because unlike the divergence, the exterior derivative commutes with a change of (curvilinear) coordinate system.

Generalizations

The divergence of a vector field can be defined in any number of dimensions. If,

$$F = (F_1, F_2, \ldots F_n),$$

in a Euclidean coordinate system with coordinates x_1, x_2, ..., x_n, define,

$$\text{div}\, F = \nabla \cdot F = \frac{\partial F_1}{\partial x_1} + \frac{\partial F_2}{\partial x_2} + \cdots + \frac{\partial F_n}{\partial x_n}.$$

The appropriate expression is more complicated in curvilinear coordinates.

In the case of one dimension, F reduces to a regular function, and the divergence reduces to the derivative.

For any n, the divergence is a linear operator, and it satisfies the "product rule",

$$\nabla \cdot (\varphi F) = (\nabla \varphi) \cdot F + \varphi (\nabla \cdot F)$$

for any scalar-valued function φ.

The divergence of a vector field extends naturally to any differentiable manifold of dimension n that has a volume form (or density) μ, e.g. a Riemannian or Lorentzian manifold. Generalising the construction of a two-form for a vector field on R³, on such a manifold a vector field X defines an $(n-1)$-form $j = i_X \mu$ obtained by contracting X with μ. The divergence is then the function defined by,

$$dj = (\text{div}\, X)\mu.$$

Standard formulas for the Lie derivative allow us to reformulate this as,

$$\mathcal{L}_X \mu = (\text{div}\, X)\mu.$$

This means that the divergence measures the rate of expansion of a volume element as we let it flow with the vector field.

On a pseudo-Riemannian manifold, the divergence with respect to the metric volume form can be computed in terms of the Levi-Civita connection ∇:

$$\text{div}\, X = \nabla \cdot X = X^a{}_{;a},$$

where the second expression is the contraction of the vector field valued 1-form ∇X with itself and the last expression is the traditional coordinate expression from Ricci calculus.

An equivalent expression without using connection is,

$$\text{div}(X) = \frac{1}{\sqrt{\det g}} \partial_a \left(\sqrt{\det g}\, X^a \right),$$

where g is the metric and ∂_a denotes the partial derivative with respect to coordinate x^a.

Divergence can also be generalised to tensors. In Einstein notation, the divergence of a contravariant vector F^μ is given by,

$$\nabla \cdot F = \nabla_\mu F^\mu,$$

where ∇_μ denotes the covariant derivative.

Equivalently, some authors define the divergence of a mixed tensor by using the musical isomorphism ♯: if T is a (p, q)-tensor (p for the contravariant vector and q for the covariant one), then we define the *divergence of T* to be the $(p, q - 1)$-tensor,

$$(\text{div}\,T)(Y_1,\ldots,Y_{q-1}) = \text{trace}\left(X \mapsto \sharp(\nabla T)(X,\cdot,Y_1,\ldots,Y_{q-1}) \right);$$

that is, we take the trace over the *first two* covariant indices of the covariant derivative.

CURL

In vector calculus, the curl is a vector operator that describes the infinitesimal rotation of a vector field in three-dimensional Euclidean space. At every point in the field, the curl of that point is represented by a vector. The attributes of this vector (length and direction) characterize the rotation at that point.

The direction of the curl is the axis of rotation, as determined by the right-hand rule, and the magnitude of the curl is the magnitude of rotation. If the vector field represents the flow velocity of a moving fluid, then the curl is the circulation density of the fluid. A vector field whose curl is zero is called irrotational. The curl is a form of differentiation for vector fields. The corresponding form of the fundamental theorem of calculus is Stokes' theorem, which relates the surface integral of the curl of a vector field to the line integral of the vector field around the boundary curve.

The alternative terminology *rotation* or *rotational* and alternative notations rot F and $\nabla \times F$ are often used (the former especially in many European countries, the latter, using the del (or nabla) operator and the cross product, is more used in other countries) for curl F.

Unlike the gradient and divergence, curl does not generalize as simply to other dimensions; some generalizations are possible, but only in three dimensions is the geometrically defined curl of a vector field again a vector field. This is a phenomenon similar to the 3-dimensional cross product, and the connection is reflected in the notation $\nabla \times$ for the curl.

The name "curl" was first suggested by James Clerk Maxwell in 1871 but the concept was apparently first used in the construction of an optical field theory by James MacCullagh in 1839.

The curl of a vector field F, denoted by curl F, or $\nabla \times$ F, or rot F, at a point is defined in terms of its projection onto various lines through the point. If $\hat{n}$ is any unit vector, the projection of the curl of F onto $\hat{n}$ is defined to be the limiting value of a closed line integral in a plane orthogonal to $\hat{n}$ divided by the area enclosed, as the path of integration is contracted around the point.

The curl operator maps continuously differentiable functions $f: R^3 \to R^3$ to continuous functions $g: R^3 \to R^3$, and more generally, it maps C^k functions in R^3 to C^{k-1} functions in R^3.

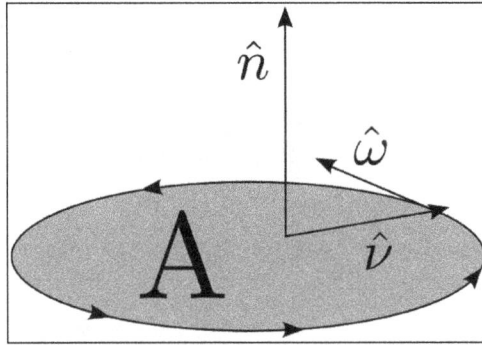

Convention for vector orientation of the line integral.

Implicitly, curl is defined:

$$(\nabla \times F)(p) \cdot \hat{n} \overset{\text{def}}{=} \lim_{A \to 0} \left(\frac{1}{|A|} \oint_C F \cdot d\mathbf{r} \right)$$

where $\oint_C F \cdot d\mathbf{r}$ is a line integral along the boundary of the area in question, and $|A|$ is the magnitude of the area. This equation defines the projection of the curl of F onto $\hat{n}$, where $\hat{n}$ is the normal vector to the surface bounded by C; and C is defined via the right-hand rule.

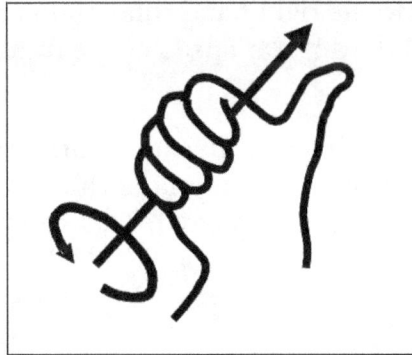

Right hand rule for curve orientation. The thumb points in the direction of $\hat{n}$, and the fingers curl along the orientation of C.

The above formula means that the curl of a vector field is defined as the infinitesimal area density of the *circulation* of that field. To this definition fit naturally:

- The Kelvin–Stokes theorem, as a global formula corresponding to the definition, and

- The following "easy to memorize" definition of the curl in curvilinear orthogonal coordinates, e.g. in Cartesian coordinates, spherical, cylindrical, or even elliptical or parabolic coordinates:

$$(\text{curl}\,F)_1 = \frac{1}{h_2 h_3}\left(\frac{\partial(h_3 F_3)}{\partial u_2} - \frac{\partial(h_2 F_2)}{\partial u_3}\right),$$

$$(\text{curl}\,F)_2 = \frac{1}{h_3 h_1}\left(\frac{\partial(h_1 F_1)}{\partial u_3} - \frac{\partial(h_3 F_3)}{\partial u_1}\right),$$

$$(\text{curl}\,F)_3 = \frac{1}{h_1 h_2}\left(\frac{\partial(h_2 F_2)}{\partial u_1} - \frac{\partial(h_1 F_1)}{\partial u_2}\right).$$

The equation for each component $(\text{curl}\,F)_k$ can be obtained by exchanging each occurrence of a subscript 1, 2, 3 in cyclic permutation: $1 \to 2$, $2 \to 3$, and $3 \to 1$ (where the subscripts represent the relevant indices).

If (x_1, x_2, x_3) are the Cartesian coordinates and (u_1, u_2, u_3) are the orthogonal coordinates then,

$$h_i = \sqrt{\left(\frac{\partial x_1}{\partial u_i}\right)^2 + \left(\frac{\partial x_2}{\partial u_i}\right)^2 + \left(\frac{\partial x_3}{\partial u_i}\right)^2}$$

is the length of the coordinate vector corresponding to u_i. The remaining two components of curl result from cyclic permutation of indices: $3,1,2 \to 1,2,3 \to 2,3,1$.

Intuitive Interpretation

Suppose the vector field describes the velocity field of a fluid flow (such as a large tank of liquid or gas) and a small ball is located within the fluid or gas (the centre of the ball being fixed at a certain point). If the ball has a rough surface, the fluid flowing past it will make it rotate. The rotation axis (oriented according to the right hand rule) points in the direction of the curl of the field at the centre of the ball, and the angular speed of the rotation is half the magnitude of the curl at this point.

The curl of the vector at any point is given by the rotation of an infinitesimal area in the xy-plane (for z-axis component of the curl), zx-plane (for y-axis component of the curl) and yz-plane (for x-axis component of the curl vector). This can be clearly seen in the examples below.

Inverse

The inverse curl of a three-dimensional vector field can be obtained up to an integration constant and an unknown irrotational field with the Biot–Savart law.

Usage

In practice, the above definition is rarely used because in virtually all cases, the curl operator can be applied using some set of curvilinear coordinates, for which simpler representations have been derived.

The notation $\nabla \times F$ has its origins in the similarities to the 3-dimensional cross product, and it is useful as a mnemonic in Cartesian coordinates if ∇ is taken as a vector differential operator del. Such notation involving operators is common in physics and algebra.

Expanded in 3-dimensional Cartesian coordinates, $\nabla \times F$ is, for F composed of $[F_x, F_y, F_z]$:

$$\begin{vmatrix} \hat{\imath} & \hat{\jmath} & \hat{k} \\ \dfrac{\partial}{\partial x} & \dfrac{\partial}{\partial y} & \dfrac{\partial}{\partial z} \\ F_x & F_y & F_z \end{vmatrix}$$

where i, j, and k are the unit vectors for the x-, y-, and z-axes, respectively. This expands as follows:

$$\left(\frac{\partial F_z}{\partial y} - \frac{\partial F_y}{\partial z} \right)\hat{y} + \left(\frac{\partial F_x}{\partial z} - \frac{\partial F_z}{\partial x} \right)\hat{\jmath} + \left(\frac{\partial F_y}{\partial x} - \frac{\partial F_x}{\partial y} \right)\hat{k} = \begin{bmatrix} \dfrac{\partial F_z}{\partial y} - \dfrac{\partial F_y}{\partial z} \\ \dfrac{\partial F_x}{\partial z} - \dfrac{\partial F_z}{\partial x} \\ \dfrac{\partial F_y}{\partial x} - \dfrac{\partial F_x}{\partial y} \end{bmatrix}$$

Although expressed in terms of coordinates, the result is invariant under proper rotations of the coordinate axes but the result inverts under reflection.

In a general coordinate system, the curl is given by,

$$(\nabla \times F)^k = \varepsilon^{k\ell m}\nabla_\ell F_m$$

where ε denotes the Levi-Civita tensor and ∇ the covariant derivative, the metric tensor is used to lower the index on F, and the Einstein summation convention implies that repeated indices are summed over. Equivalently,

$$(\nabla \times F) = e_k\,\varepsilon^{k\ell m}\nabla_l F_m$$

where e_k are the coordinate vector fields. Equivalently, using the exterior derivative, the curl can be expressed as:

$$\nabla \times F = \left(\star\left(dF^\flat \right) \right)^\sharp$$

Here $\flat$ and $\sharp$ are the musical isomorphisms, and $\star$ is the Hodge star operator. This formula shows how to calculate the curl of F in any coordinate system, and how to extend the curl to any oriented three-dimensional Riemannian manifold. Since this depends on a choice of orientation, curl is a chiral operation. In other words, if the orientation is reversed, then the direction of the curl is also reversed.

Example: Take the vector field:

$$F(x, y, z) = y\hat{i} - x\hat{j}.$$

For clarity, this can be decomposed as follows:

$$F_x = y, F_y = -x, F_z = 0.$$

Its corresponding plot:

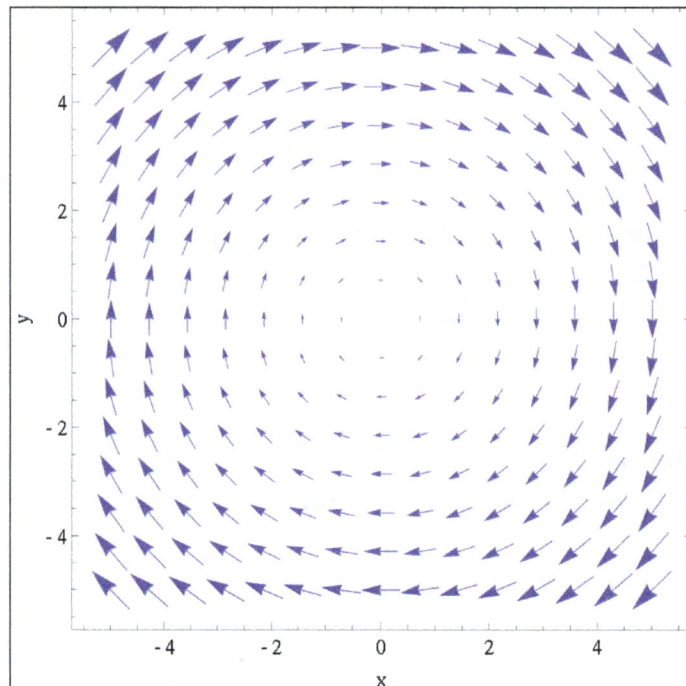

Upon visual inspection, the field can be described as "rotating". If a stationary object were to be placed in the field with the vectors representing a linear force, the object would rotate clockwise.

Calculating the curl:

$$\nabla \times F = 0\hat{i} + 0\hat{j} + \left(\frac{\partial}{\partial x}(-x) - \frac{\partial}{\partial y} y \right) \hat{k} = -2\hat{k}$$

The resulting vector field describing the curl would be uniformly going in the negative z direction. The results of this equation align with what could have been predicted using the right-hand rule using a right-handed coordinate system. Being a uniform vector field, the object described before would have the same rotational intensity regardless of where it was placed.

The plot describing the curl of F:

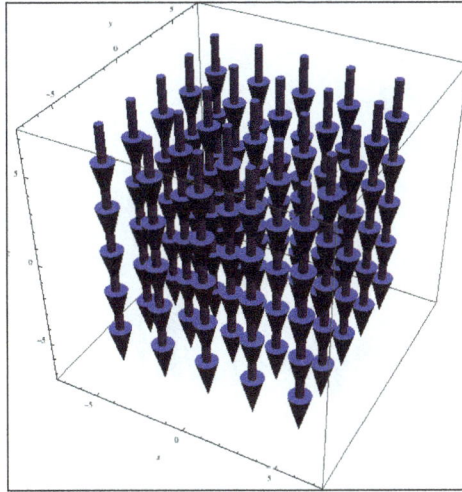

Example: Take the vector field:

$$F(x, y, z) = -x^2 \hat{j}.$$

Its corresponding plot:

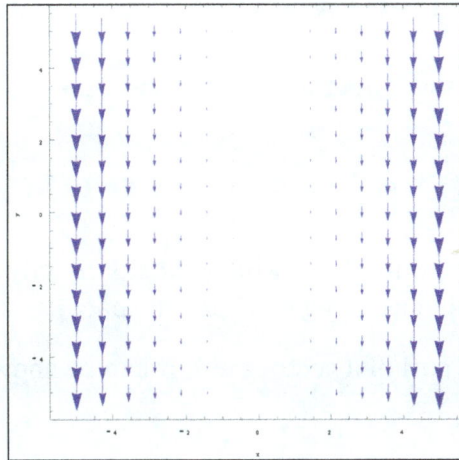

Upon initial inspection, curl existing in this graph would not be obvious. However, taking the object in the previous example, and placing it anywhere on the line $x = 3$, the force exerted on the right side would be slightly greater than the force exerted on the left, causing it to rotate clockwise. Using the right-hand rule, it can be predicted that the resulting curl would be straight in the negative z direction. Inversely, if placed on $x = -3$, the object would rotate counterclockwise and the right-hand rule would result in a positive z direction.

Calculating the curl:

$$\nabla \times F = 0\hat{i} + 0\hat{j} + \frac{\partial}{\partial x}(-x^2)\hat{k} = -2x\hat{k}.$$

As predicted, the curl points in the negative z direction when x is positive and vice versa. In this field, the intensity of rotation would be greater as the object moves away from the plane $x = 0$.

The plot describing the curl of F:

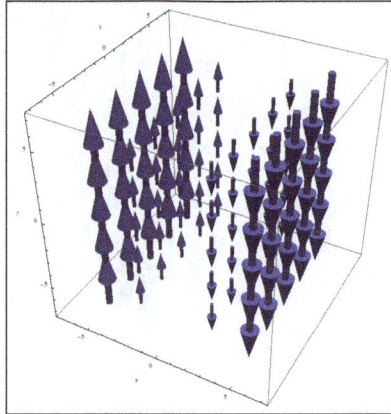

Identities

In general curvilinear coordinates (not only in Cartesian coordinates), the curl of a cross product of vector fields v and F can be shown to be,

$$\nabla \times (v \times F) = ((\nabla \cdot F) + F \cdot \nabla) v - ((\nabla \cdot v) + v \cdot \nabla) F.$$

Interchanging the vector field v and ∇ operator, we arrive at the cross product of a vector field with curl of a vector field:

$$v \times (\nabla \times F) = \nabla_F (v \cdot F) - (v \cdot \nabla) F,$$

where ∇_F is the Feynman subscript notation, which considers only the variation due to the vector field F (i.e., in this case, v is treated as being constant in space).

Another example is the curl of a curl of a vector field. It can be shown that in general coordinates,

$$\nabla \times (\nabla \times F) = \nabla(\nabla \cdot F) - \nabla^2 F,$$

and this identity defines the vector Laplacian of F, symbolized as $\nabla^2 F$.

The curl of the gradient of *any* scalar field φ is always the zero vector field,

$$\nabla \times (\nabla \varphi) = 0$$

which follows from the antisymmetry in the definition of the curl, and the symmetry of second derivatives.

If φ is a scalar valued function and F is a vector field then,

$$\nabla \times (\varphi F) = \nabla \varphi \times F + \varphi \nabla \times F$$

Descriptive Examples:

- In a vector field describing the linear velocities of each part of a rotating disk, the curl has the same value at all points.

- Of the four Maxwell's equations, two—Faraday's law and Ampère's law—can be compactly expressed using curl. Faraday's law states that the curl of an electric field is equal to the opposite of the time rate of change of the magnetic field, while Ampère's law relates the curl of the magnetic field to the current and rate of change of the electric field.

Generalizations

The vector calculus operations of grad, curl, and div are most easily generalized and understood in the context of differential forms, which involves a number of steps. In a nutshell, they correspond to the derivatives of 0-forms, 1-forms, and 2-forms, respectively. The geometric interpretation of curl as rotation corresponds to identifying bivectors (2-vectors) in 3 dimensions with the special orthogonal Lie algebra $\mathfrak{so}$ (3) of infinitesimal rotations (in coordinates, skew-symmetric 3 × 3 matrices), while representing rotations by vectors corresponds to identifying 1-vectors (equivalently, 2-vectors) and $\mathfrak{so}$ (3), these all being 3-dimensional spaces.

Differential Forms

In 3 dimensions, a differential 0-form is simply a function $f(x, y, z)$; a differential 1-form is the following expression:

$$a_1\,dx + a_2\,dy + a_3\,dz;$$

a differential 2-form is the formal sum:

$$a_{12}\,dx \wedge dy + a_{13}\,dx \wedge dz + a_{23}\,dy \wedge dz;$$

and a differential 3-form is defined by a single term:

$$a_{123}\,dx \wedge dy \wedge dz.$$

(Here the a-coefficients are real functions; the "wedge products", e.g. $dx \wedge dy$, can be interpreted as some kind of oriented area elements, $dx \wedge dy = -dy \wedge dx$, etc.)

The exterior derivative of a k-form in $\mathbb{R}^3$ is defined as the $(k + 1)$-form from above—and in $\mathbb{R}^n$ if, e.g.,

$$\omega^{(k)} = \sum_{\substack{i_1 < i_2 < \cdots < i_k \\ \forall i_\nu \in 1, \ldots, n}} a_{i_1, \ldots, i_k}\, dx_{i_1} \wedge \cdots \wedge dx_{i_k},$$

then the exterior derivative d leads to,

$$d\omega^{(k)} = \sum_{\substack{j=1 \\ i_1 < \cdots < i_k}}^{n} \frac{\partial a_{i_1, \ldots, i_k}}{\partial x_j} dx_j \wedge dx_{i_1} \wedge \cdots \wedge dx_{i_k}.$$

The exterior derivative of a 1-form is therefore a 2-form, and that of a 2-form is a 3-form. On the other hand, because of the interchangeability of mixed derivatives, e.g. because of,

$$\frac{\partial^2}{\partial x \partial y} = \frac{\partial^2}{\partial y \partial x},$$

the two fold application of the exterior derivative leads to 0.

Thus, denoting the space of k-forms by $\Omega^k(\mathbb{R}^3)$ and the exterior derivative by d one gets a sequence:

$$0 \xrightarrow{d} \Omega^0(\mathbb{R}^3) \xrightarrow{d} \Omega^1(\mathbb{R}^3) \xrightarrow{d} \Omega^2(\mathbb{R}^3) \xrightarrow{d} \Omega^3(\mathbb{R}^3) \xrightarrow{d} 0.$$

Here $\Omega^k(\mathbb{R}^n)$ is the space of sections of the exterior algebra $\Lambda^k(\mathbb{R}^n)$ vector bundle over $\mathbb{R}^n$, whose dimension is the binomial coefficient $\binom{n}{k}$; note that $\Omega^k(\mathbb{R}^3) = 0$ for $k > 3$ or $k < 0$. Writing only dimensions, one obtains a row of Pascal's triangle:

$$0 \to 1 \to 3 \to 3 \to 1 \to 0;$$

the 1-dimensional fibers correspond to scalar fields, and the 3-dimensional fibers to vector fields, as described below. Modulo suitable identifications, the three nontrivial occurrences of the exterior derivative correspond to grad, curl, and div.

Differential forms and the differential can be defined on any Euclidean space, or indeed any manifold, without any notion of a Riemannian metric. On a Riemannian manifold, or more generally pseudo-Riemannian manifold, k-forms can be identified with k-vector fields (k-forms are k-covector fields, and a pseudo-Riemannian metric gives an isomorphism between vectors and covectors), and on an *oriented* vector space with a nondegenerate form (an isomorphism between vectors and covectors), there is an isomorphism between k-vectors and $(n - k)$-vectors; in particular on (the tangent space of) an oriented pseudo-Riemannian manifold. Thus on an oriented pseudo-Riemannian manifold, one can interchange k-forms, k-vector fields, $(n - k)$-forms, and $(n - k)$-vector fields; this is known as Hodge duality. Concretely, on $\mathbb{R}^3$ this is given by:

- 1-forms and 1-vector fields: the 1-form $a_x \, dx + a_y \, dy + a_z \, dz$ corresponds to the vector field (a_x, a_y, a_z).

- 1-forms and 2-forms: one replaces dx by the dual quantity $dy \wedge dz$ (i.e., omit dx), and likewise, taking care of orientation: dy corresponds to $dz \wedge dx = -dx \wedge dz$, and dz corresponds to $dx \wedge dy$. Thus the form $a_x \, dx + a_y \, dy + a_z \, dz$ corresponds to the "dual form" $a_z \, dx \wedge dy + a_y \, dz \wedge dx + a_x \, dy \wedge dz$.

Thus, identifying 0-forms and 3-forms with scalar fields, and 1-forms and 2-forms with vector fields:

- Grad takes a scalar field (0-form) to a vector field (1-form);

- Curl takes a vector field (1-form) to a pseudovector field (2-form);

- Div takes a pseudovector field (2-form) to a pseudoscalar field (3-form).

On the other hand, the fact that $d^2 = 0$ corresponds to the identities,

$$\nabla \times (\nabla f) = 0$$

for any scalar field f, and

$$\nabla \cdot (\nabla \times v) = 0$$

for any vector field v.

Grad and div generalize to all oriented pseudo-Riemannian manifolds, with the same geometric interpretation, because the spaces of 0-forms and n-forms is always (fiberwise) 1-dimensional and can be identified with scalar fields, while the spaces of 1-forms and $(n - 1)$-forms are always fiberwise n-dimensional and can be identified with vector fields.

Curl does not generalize in this way to 4 or more dimensions (or down to 2 or fewer dimensions); in 4 dimensions the dimensions are:

$$0 \to 1 \to 4 \to 6 \to 4 \to 1 \to 0;$$

so the curl of a 1-vector field (fiberwise 4-dimensional) is a *2-vector field*, which is fiberwise 6-dimensional one has,

$$\omega^{(2)} = \sum_{i<k=1,2,3,4} a_{i,k}\, dx_i \wedge dx_k,$$

which yields a sum of six independent terms, and cannot be identified with a 1-vector field. Nor can one meaningfully go from a 1-vector field to a 2-vector field to a 3-vector field ($4 \to 6 \to 4$), as taking the differential twice yields zero ($d^2 = 0$). Thus there is no curl function from vector fields to vector fields in other dimensions arising in this way.

However, one can define a curl of a vector field as a *2-vector field* in general.

Curl Geometrically

2-vectors correspond to the exterior power $\Lambda^2 V$; in the presence of an inner product, in coordinates these are the skew-symmetric matrices, which are geometrically considered as the special orthogonal Lie algebra so (V) of infinitesimal rotations. This has $\binom{n}{2} = \frac{1}{2}n(n-1)$ dimensions, and allows one to interpret the differential of a 1-vector field as its infinitesimal rotations. Only in 3 dimensions (or trivially in 0 dimensions) does $n = \frac{1}{2}n(n-1)$, which is the most elegant and common case. In 2 dimensions the curl of a vector field is not a vector field but a function, as 2-dimensional rotations are given by an angle (a scalar – an orientation is required to choose whether one counts clockwise or counterclockwise rotations as positive); this is not the div, but is rather perpendicular to it. In 3 dimensions the curl of a vector field is a vector field as is familiar (in 1 and 0 dimensions the curl of a vector field is 0, because there are no non-trivial 2-vectors), while in 4 dimensions the curl of a vector field is, geometrically, at each point an element of the 6-dimensional Lie algebra so (4).

Note also that the curl of a 3-dimensional vector field which only depends on 2 coordinates (say x and y) is simply a vertical vector field (in the z direction) whose magnitude is the curl of the 2-dimensional vector field.

Considering curl as a 2-vector field (an antisymmetric 2-tensor) has been used to generalize vector calculus and associated physics to higher dimensions.

VECTOR FIELDS IN CYLINDRICAL AND SPHERICAL COORDINATES

Cylindrical Coordinate System

Vector Fields

Vectors are defined in cylindrical coordinates by (ρ, φ, z), where

- ρ is the length of the vector projected onto the xy-plane,

- φ is the angle between the projection of the vector onto the xy-plane (i.e. ρ) and the positive x-axis ($0 \leq \varphi < 2\pi$),

- z is the regular z-coordinate.

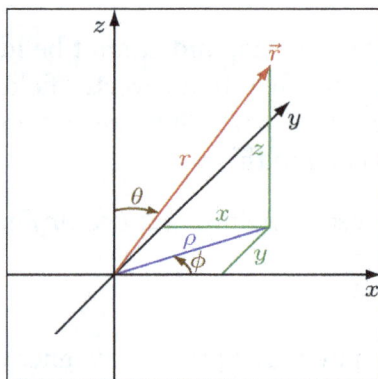

(ρ, φ, z) is given in cartesian coordinates by:

$$\begin{bmatrix} \rho \\ \phi \\ z \end{bmatrix} = \begin{bmatrix} \sqrt{x^2 + y^2} \\ \arctan(y/x) \\ z \end{bmatrix}, \quad 0 \leq \phi < 2\pi,$$

or inversely by:

$$\begin{bmatrix} x \\ y \\ z \end{bmatrix} = \begin{bmatrix} \rho \cos\phi \\ \rho \sin\phi \\ z \end{bmatrix}.$$

Any vector field can be written in terms of the unit vectors as:

$$A = A_x \hat{x} + A_y \hat{y} + A_z \hat{z} = A_\rho \hat{\rho} + A_\phi \hat{\phi} + A_z \hat{z}$$

The cylindrical unit vectors are related to the cartesian unit vectors by:

$$\begin{bmatrix} \hat{\rho} \\ \hat{\phi} \\ \hat{z} \end{bmatrix} = \begin{bmatrix} \cos\phi & \sin\phi & 0 \\ -\sin\phi & \cos\phi & 0 \\ 0 & 0 & 1 \end{bmatrix} \begin{bmatrix} \hat{x} \\ \hat{y} \\ \hat{z} \end{bmatrix}$$

Time Derivative of a Vector Field

To find out how the vector field A changes in time we calculate the time derivatives. For this purpose we use Newton's notation for the time derivative ($\dot{A}$). In cartesian coordinates this is simply:

$$\dot{A} = \dot{A}_x \hat{x} + \dot{A}_y \hat{y} + \dot{A}_z \hat{z}$$

However, in cylindrical coordinates this becomes:

$$\dot{A} = \dot{A}_\rho \hat{\rho} + A_\rho \dot{\hat{\rho}} + \dot{A}_\phi \hat{\phi} + A_\phi \dot{\hat{\phi}} + \dot{A}_z \hat{z} + A_z \dot{\hat{z}}$$

We need the time derivatives of the unit vectors. They are given by:

$$\dot{\hat{\rho}} = \dot{\phi}\hat{\phi}$$

$$\dot{\hat{\phi}} = -\dot{\phi}\hat{\rho}$$

$$\dot{\hat{z}} = 0$$

So the time derivative simplifies to:

$$\dot{A} = \hat{\rho}(\dot{A}_\rho - A_\phi \dot{\phi}) + \hat{\phi}(\dot{A}_\phi + A_\rho \dot{\phi}) + \hat{z}\dot{A}_z$$

Second Time Derivative of a Vector Field

The second time derivative is of interest in physics, as it is found in equations of motion for classical mechanical systems. The second time derivative of a vector field in cylindrical coordinates is given by:

$$\ddot{A} = \hat{\rho}(\ddot{A}_\rho - A_\phi \ddot{\phi} - 2\dot{A}_\phi \dot{\phi} - A_\rho \dot{\phi}^2) + \hat{\phi}(\ddot{A}_\phi + A_\rho \ddot{\phi} + 2\dot{A}_\rho \dot{\phi} - A_\phi \dot{\phi}^2) + \hat{z}\ddot{A}_z$$

To understand this expression, we substitute A = P, where p is the vector (\rho, θ, z).

This means that $A = P = \rho\hat{\rho} + z\hat{z}$.

After substituting we get:

$$\ddot{P} = \hat{\rho}(\ddot{\rho} - \rho\dot{\phi}^2) + \hat{\phi}(\rho\ddot{\phi} + 2\dot{\rho}\dot{\phi}) + \hat{z}\ddot{z}$$

In mechanics, the terms of this expression are called:

$$\ddot{\rho}\hat{\rho} = \text{Central outward acceleration}$$

$$-\rho\dot{\phi}^2\hat{\rho} = \text{Centripetal acceleration}$$

$$\rho\ddot{\phi}\hat{\phi} = \text{Angular acceleration}$$

$$2\dot{\rho}\dot{\phi}\hat{\phi} = \text{Coriolis effect}$$

$$\ddot{z}\hat{z} = \text{Z-acceleration}$$

Spherical Coordinate System

Vector Fields

Vectors are defined in spherical coordinates by (r, θ, φ), where

- r is the length of the vector,

- θ is the angle between the positive Z-axis and the vector in question ($0 \le \theta \le \pi$), and

- φ is the angle between the projection of the vector onto the X-Y-plane and the positive X-axis ($0 \le \varphi < 2\pi$).

(r, θ, φ) is given in Cartesian coordinates by:

$$\begin{bmatrix} r \\ \theta \\ \phi \end{bmatrix} = \begin{bmatrix} \sqrt{x^2 + y^2 + z^2} \\ \arccos(z/r) \\ \arctan(y/x) \end{bmatrix}, \quad 0 \le \theta \le \pi, \quad 0 \le \phi < 2\pi,$$

or inversely by:

$$\begin{bmatrix} x \\ y \\ z \end{bmatrix} = \begin{bmatrix} r\sin\theta\cos\phi \\ r\sin\theta\sin\phi \\ r\cos\theta \end{bmatrix}.$$

Any vector field can be written in terms of the unit vectors as:

$$A = A_x\hat{x} + A_y\hat{y} + A_z\hat{z} = A_r\hat{r} + A_\theta\hat{\theta} + A_\phi\hat{\phi}$$

The spherical unit vectors are related to the cartesian unit vectors by:

$$\begin{bmatrix} \hat{r} \\ \hat{\theta} \\ \hat{\phi} \end{bmatrix} = \begin{bmatrix} \sin\theta\cos\phi & \sin\theta\sin\phi & \cos\theta \\ \cos\theta\cos\phi & \cos\theta\sin\phi & -\sin\theta \\ -\sin\phi & \cos\phi & 0 \end{bmatrix}\begin{bmatrix} \hat{x} \\ \hat{y} \\ \hat{z} \end{bmatrix}$$

The matrix is an orthogonal matrix, that is, its inverse is simply its transpose.

So the cartesian unit vectors are related to the spherical unit vectors by:

$$\begin{bmatrix} \hat{x} \\ \hat{y} \\ \hat{z} \end{bmatrix} = \begin{bmatrix} \sin\theta\cos\phi & \cos\theta\cos\phi & -\sin\phi \\ \sin\theta\sin\phi & \cos\theta\sin\phi & \cos\phi \\ \cos\theta & -\sin\theta & 0 \end{bmatrix} \begin{bmatrix} \hat{r} \\ \hat{\theta} \\ \hat{\phi} \end{bmatrix}$$

Time Derivative of a Vector Field

To find out how the vector field A changes in time we calculate the time derivatives. In cartesian coordinates this is simply:

$$\dot{A} = \dot{A}_x\,\hat{x} + \dot{A}_y\,\hat{y} + \dot{A}_z\,\hat{z}$$

However, in spherical coordinates this becomes:

$$\dot{A} = \dot{A}_r\hat{r} + A_r\dot{\hat{r}} + \dot{A}_\theta\,\hat{\theta} + A_\theta\dot{\hat{\theta}} + \dot{A}_\phi\,\hat{\phi} + A_\phi\dot{\hat{\phi}}$$

We need the time derivatives of the unit vectors. They are given by:

$$\dot{\hat{r}} = \dot{\theta}\hat{\theta} + \dot{\phi}\sin\theta\hat{\phi}$$

$$\dot{\hat{\theta}} = -\dot{\theta}\,\hat{r} + \dot{\phi}\cos\theta\hat{\phi}$$

$$\dot{\hat{\phi}} = -\dot{\phi}\sin\theta\,\hat{r} - \dot{\phi}\cos\theta\hat{\theta}$$

So the time derivative becomes:

$$\dot{A} = \hat{r}(\dot{A}_r - A_\theta\dot{\theta} - A_\phi\dot{\phi}\sin\theta) + \hat{\theta}(\dot{A}_\theta + A_r\dot{\theta} - A_\phi\dot{\phi}\cos\theta) + \hat{\phi}(\dot{A}_\phi + A_r\dot{\phi}\sin\theta + A_\theta\dot{\phi}\cos\theta)$$

VECTOR SPHERICAL HARMONICS

In mathematics, vector spherical harmonics (VSH) are an extension of the scalar spherical harmonics for use with vector fields. The components of the VSH are complex-valued functions expressed in the spherical coordinate basis vectors.

Several conventions have been used to define the VSH. Given a scalar spherical harmonic $Y_{\ell m}(\theta, \varphi)$, we define three VSH:

- $Y_{lm} = Y_{lm}\,\hat{r}$,

- $\Psi_{lm} = r\nabla Y_{lm}$,

- $\Phi_{lm} = r \times \nabla Y_{lm}$,

with $\hat{r}$ being the unit vector along the radial direction in spherical coordinates and r the vector along the radial direction with the same norm as the radius, i.e., $r = r\hat{r}$. The radial factors are included to guarantee that the dimensions of the VSH are the same as those of the ordinary spherical harmonics and that the VSH do not depend on the radial spherical coordinate.

The interest of these new vector fields is to separate the radial dependence from the angular one when using spherical coordinates, so that a vector field admits a multipole expansion:

$$E = \sum_{l=0}^{\infty} \sum_{m=-l}^{l} \left(E_{lm}^{r}(r)\, Y_{lm} + E_{lm}^{(1)}(r)\Psi_{lm} + E_{lm}^{(2)}(r)\Phi_{lm} \right).$$

The labels on the components reflect that E_{lm}^{r} is the radial component of the vector field, while $E_{lm}^{(1)}$ and $E_{lm}^{(2)}$ are transverse components (with respect to the radius vector **r**).

Main Properties

Symmetry

Like the scalar spherical harmonics, the VSH satisfy,

$$Y_{l,-m} = (-1)^{m}\, Y_{lm}^{*},$$
$$\Psi_{l,-m} = (-1)^{m}\, \Psi_{lm}^{*},$$
$$\Phi_{l,-m} = (-1)^{m}\, \Phi_{lm}^{*},$$

which cuts the number of independent functions roughly in half. The star indicates complex conjugation.

Orthogonality

The VSH are orthogonal in the usual three-dimensional way at each point:

$$Y_{lm}(r)\cdot\Psi_{lm}(r) = 0,$$
$$Y_{lm}(r)\cdot\Phi_{lm}(r) = 0,$$
$$\Psi_{lm}(r)\cdot\Phi_{lm}(r) = 0.$$

They are also orthogonal in Hilbert space:

$$\int Y_{lm}\cdot Y_{l'm'}^{*}\, d\Omega = \delta_{ll'}\delta_{mm'},$$
$$\int \Psi_{lm}\cdot\Psi_{l'm'}^{*}\, d\Omega = l(l+1)\delta_{ll'}\delta_{mm'},$$
$$\int \Phi_{lm}\cdot\Phi_{l'm'}^{*}\, d\Omega = l(l+1)\delta_{ll'}\delta_{mm'},$$
$$\int Y_{lm}\cdot\Psi_{l'm'}^{*}\, d\Omega = 0,$$
$$\int Y_{lm}\cdot\Phi_{l'm'}^{*}\, d\Omega = 0,$$
$$\int \Psi_{lm}\cdot\Phi_{l'm'}^{*}\, d\Omega = 0.$$

An additional result at a single point r is, for all l, m, l', m',

$$Y_{lm}(r) \cdot \Psi_{l'm'}(r) = 0,$$
$$Y_{lm}(r) \cdot \Phi_{l'm'}(r) = 0.$$

Vector Multipole Moments

The orthogonality relations allow one to compute the spherical multipole moments of a vector field as:

$$E_{lm}^r = \int E \cdot Y_{lm}^* \, d\Omega,$$

$$E_{lm}^{(1)} = \frac{1}{l(l+1)} \int E \cdot \Psi_{lm}^* \, d\Omega,$$

$$E_{lm}^{(2)} = \frac{1}{l(l+1)} \int E \cdot \Phi_{lm}^* \, d\Omega.$$

The Gradient of a Scalar Field

Given the multipole expansion of a scalar field,

$$\phi = \sum_{l=0}^{\infty} \sum_{m=-l}^{l} \phi_{lm}(r) Y_{lm}(\theta, \phi),$$

we can express its gradient in terms of the VSH as,

$$\nabla \phi = \sum_{l=0}^{\infty} \sum_{m=-l}^{l} \left(\frac{d\phi_{lm}}{dr} Y_{lm} + \frac{\phi_{lm}}{r} \Psi_{lm} \right).$$

Divergence

For any multipole field we have,

$$\nabla \cdot \left(f(r) Y_{lm} \right) = \left(\frac{df}{dr} + \frac{2}{r} f \right) Y_{lm},$$

$$\nabla \cdot \left(f(r) \Psi_{lm} \right) = -\frac{l(l+1)}{r} f Y_{lm},$$

$$\nabla \cdot \left(f(r) \Phi_{lm} \right) = 0.$$

By superposition we obtain the divergence of any vector field:

$$\nabla \cdot E = \sum_{l=0}^{\infty} \sum_{m=-l}^{l} \left(\frac{dE_{lm}^r}{dr} + \frac{2}{r} E_{lm}^r - \frac{l(l+1)}{r} E_{lm}^{(1)} \right) Y_{lm}.$$

We see that the component on $\Phi_{\ell m}$ is always solenoidal.

Curl

For any multipole field we have,

$$\nabla \times \left(f(r) \, \mathrm{Y}_{lm} \right) = -\frac{1}{r} f \Phi_{lm},$$

$$\nabla \times \left(f(r) \Psi_{lm} \right) = \left(\frac{df}{dr} + \frac{1}{r} f \right) \Phi_{lm},$$

$$\nabla \times \left(f(r) \Phi_{lm} \right) = -\frac{l(l+1)}{r} f \mathrm{Y}_{lm} - \left(\frac{df}{dr} + \frac{1}{r} f \right) \Psi_{lm}.$$

By superposition we obtain the curl of any vector field:

$$\nabla \times \mathrm{E} = \sum_{l=0}^{\infty} \sum_{m=-l}^{l} \left(-\frac{l(l+1)}{r} E_{lm}^{(2)} \mathrm{Y}_{lm} - \left(\frac{dE_{lm}^{(2)}}{dr} + \frac{1}{r} E_{lm}^{(2)} \right) \Psi_{lm} + \left(-\frac{1}{r} E_{lm}^{r} + \frac{dE_{lm}^{(1)}}{dr} + \frac{1}{r} E_{lm}^{(1)} \right) \Phi_{lm} \right).$$

Laplacian

The action of the Laplace operator $\Delta = \nabla \cdot \nabla$ separates as follows:

$$\Delta \left(f(r) Z_{lm} \right) = \left(\frac{1}{r^2} \frac{\partial}{\partial r} r^2 \frac{\partial f}{\partial r} \right) Z_{lm} + f(r) \Delta Z_{lm},$$

where $Z_{lm} = \mathrm{Y}_{lm}, \Psi_{lm}, \Phi_{lm}$ and

$$\Delta \mathrm{Y}_{lm} = -\frac{1}{r^2} (2 + l(l+1)) \mathrm{Y}_{lm} + \frac{2}{r^2} \Psi_{lm},$$

$$\Delta \Psi_{lm} = \frac{2}{r^2} l(l+1) \mathrm{Y}_{lm} - \frac{1}{r^2} l(l+1) \Psi_{lm},$$

$$\Delta \Phi_{lm} = -\frac{1}{r^2} l(l+1) \Phi_{lm}.$$

Also note that this action becomes symmetric, i.e. the off-diagonal coefficients are equal to $\frac{2}{r^2} \sqrt{l(l+1)}$, for properly normalized VSH.

Examples:

First Vector Spherical Harmonics

- $l = 0$.

$$\mathrm{Y}_{00} = \sqrt{\frac{1}{4\pi}} \, \hat{\mathrm{r}},$$

$$\Psi_{00} = 0,$$

$$\Phi_{00} = 0.$$

- $l = 1$.

$$Y_{10} = \sqrt{\frac{3}{4\pi}} \cos\theta \hat{r},$$

$$Y_{11} = -\sqrt{\frac{3}{8\pi}} e^{i\varphi} \sin\theta \hat{r},$$

$$\Psi_{10} = -\sqrt{\frac{3}{4\pi}} \sin\theta \hat{\theta},$$

$$\Psi_{11} = -\sqrt{\frac{3}{8\pi}} e^{i\varphi} \left(\cos\theta \hat{\theta} + i\hat{\varphi} \right),$$

$$\Phi_{10} = -\sqrt{\frac{3}{4\pi}} \sin\theta \hat{\varphi},$$

$$\Phi_{11} = \sqrt{\frac{3}{8\pi}} e^{i\varphi} \left(i\hat{\theta} - \cos\theta \hat{\varphi} \right).$$

- $l = 2$.

$$Y_{20} = \frac{1}{4}\sqrt{\frac{5}{\pi}} (3\cos^2\theta - 1)\hat{r},$$

$$Y_{21} = -\sqrt{\frac{15}{8\pi}} \sin\theta \cos\theta e^{i\varphi} \hat{r},$$

$$Y_{22} = \frac{1}{4}\sqrt{\frac{15}{2\pi}} \sin^2\theta e^{2i\varphi} \hat{r}.$$

$$\Psi_{20} = -\frac{3}{2}\sqrt{\frac{5}{\pi}} \sin\theta \cos\theta \hat{\theta},$$

$$\Psi_{21} = -\sqrt{\frac{15}{8\pi}} e^{i\varphi} \left(\cos 2\theta \hat{\theta} + i\cos\theta \hat{\varphi} \right),$$

$$\Psi_{22} = \sqrt{\frac{15}{8\pi}} \sin\theta e^{2i\varphi} \left(\cos\theta \hat{\theta} + i\hat{\varphi} \right).$$

$$\Phi_{20} = -\frac{3}{2}\sqrt{\frac{5}{\pi}} \sin\theta \cos\theta \hat{\varphi},$$

$$\Phi_{21} = \sqrt{\frac{15}{8\pi}} e^{i\varphi} \left(i\cos\theta \hat{\theta} - \cos 2\theta \hat{\varphi} \right),$$

$$\Phi_{22} = \sqrt{\frac{15}{8\pi}} \sin\theta e^{2i\varphi} \left(-i\hat{\theta} + \cos\theta \hat{\varphi} \right).$$

Expressions for negative values of m are obtained by applying the symmetry relations.

Applications

Electrodynamics

The VSH are especially useful in the study of multipole radiation fields. For instance, a magnetic multipole is due to an oscillating current with angular frequency ω and complex amplitude:

$$\hat{J} = J(r)\Phi_{lm},$$

and the corresponding electric and magnetic fields, can be written as:

$$\hat{E} = E(r)\Phi_{lm},$$
$$\hat{B} = B^r(r)Y_{lm} + B^{(1)}(r)\Psi_{lm}.$$

Substituting into Maxwell equations, Gauss' law is automatically satisfied:

$$\nabla \cdot \hat{E} = 0,$$

while Faraday's law decouples as,

$$\nabla \times \hat{E} = -i\omega\hat{B} \quad \Rightarrow \quad \begin{cases} \dfrac{l(l+1)}{r}E = i\omega B^r, \\ \dfrac{dE}{dr} + \dfrac{E}{r} = i\omega B^{(1)}. \end{cases}$$

Gauss' law for the magnetic field implies,

$$\nabla \cdot \hat{B} = 0 \quad \Rightarrow \quad \frac{dB^r}{dr} + \frac{2}{r}B^r - \frac{l(l+1)}{r}B^{(1)} = 0,$$

and Ampère-Maxwell's equation gives,

$$\nabla \times \hat{B} = \mu_0\hat{J} + i\mu_0\varepsilon_0\omega\hat{E} \quad \Rightarrow \quad -\frac{B^r}{r} + \frac{dB^{(1)}}{dr} + \frac{B^{(1)}}{r} = \mu_0 J + i\omega\mu_0\varepsilon_0 E.$$

In this way, the partial differential equations have been transformed into a set of ordinary differential equations.

Fluid Dynamics

In the calculation of the Stokes' law for the drag that a viscous fluid exerts on a small spherical particle, the velocity distribution obeys Navier-Stokes equations neglecting inertia, i.e.

$$\nabla \cdot v = 0,$$
$$0 = -\nabla p + \eta \nabla^2 v,$$

with the boundary conditions,

$$v = 0 \quad (r = a),$$
$$v = -U_0 \quad (r \to \infty).$$

where U is the relative velocity of the particle to the fluid far from the particle. In spherical coordinates this velocity at infinity can be written as,

$$U_0 = U_0 \left(\cos \theta \hat{r} - \sin \theta \hat{\theta} \right) = U_0 \left(Y_{10} + \Psi_{10} \right).$$

The last expression suggests an expansion in spherical harmonics for the liquid velocity and the pressure,

$$p = p(r)Y_{10},$$
$$v = v^r(r) Y_{10} + v^{(1)}(r) \Psi_{10}.$$

Substitution in the Navier–Stokes equations produces a set of ordinary differential equations for the coefficients.

VECTOR-VALUED FUNCTION

A vector valued function is a function which takes a real number, t, as an input and returns a vector as an output, $\langle h(t), g(t), f(t) \rangle$.

Vector valued functions can measure many things such as position, velocity, acceleration, force, etc.

The Position vector:

$$\vec{r}(t) = \langle f(t), g(t), h(t) \rangle$$

The vector formula for a line is an example of a vector valued function.

As with lines, vector valued functions can also be written as parametric equations.

Position Vectors

As with lines, the position vector points from the origin to a location in 2-D or 3-D space.

A *position vector function* is then a function that takes a real number tt (thought of as time and called the parameter) and outputs a position vector:

$$\vec{r}(t) = \langle f(t), g(t), h(t) \rangle$$

Plotting the points at the tip of the position vector for various values of the parameter tt gives a *curve* in 2-D or 3-D space:

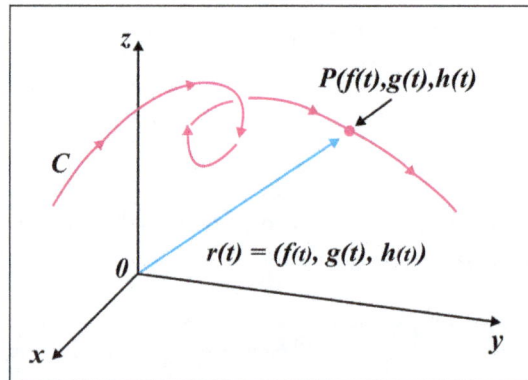

The arrows on the curve show the direction the point moves along the curve as tt increases.

Parametric Equations

Parametric equations are an alternate way to write the position function that describes a curve. The curve is written as three separate functions called the *parametric equations of the curve*.

$$x = f(t)$$
$$y = g(t)$$
$$z = h(t)$$

As with lines, there are multiple parameterizations for a curve (different position vector functions that describe the same curve).

Examples:

Lines

The Line that starts at the point P(1,3,−2) and moves towards the point Q(2,−1,3).

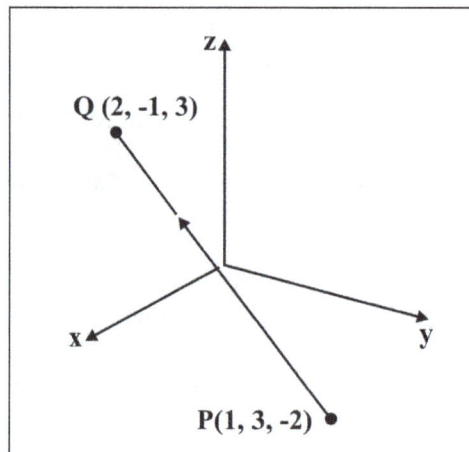

The line has a direction vector $\overrightarrow{PQ} = \langle 2-1, -1-3, 3-(-2) \rangle = \langle 1, -4, 5 \rangle$

So the vector form of the line is,

$$\begin{aligned} \vec{r}(t) &= \vec{r}_0 + tv \\ &= \langle 1, 3, -2 \rangle + t \langle 1, -4, 5 \rangle \\ &= \langle 1+t, 3-4t, -2+5t \rangle \end{aligned}$$

And the parametric equations for this line are:

$$x = 1+t$$
$$y = 3-4t$$
$$z = -2+5t$$

Parabola

Graph the vector valued function $\vec{r}(t) = \langle t^2 - t, t+1 \rangle$

Some position values at different values of t are

t	$\vec{r}$
-1	$\langle 2, 0 \rangle$
0	$\langle 0, 1 \rangle$
1	$\langle 0, 2 \rangle$
2	$\langle 2, 3 \rangle$
3	$\langle 6, 4 \rangle$

The graph of this curve is

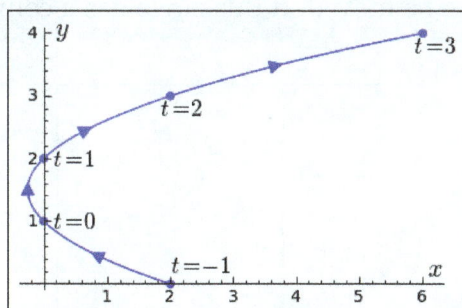

The location at different times t are labeled on the graph and the arrow show the direction of movement as t increases.

Circles in 2-D

Circles can be described using Polar Coordinates.

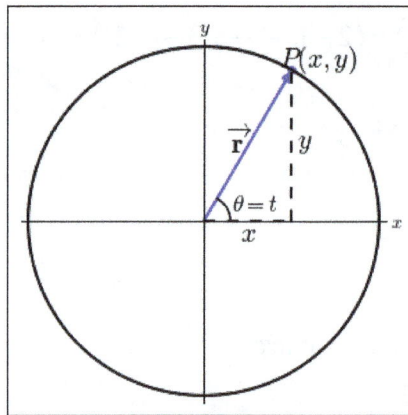

Where the parameter is the angle, t=θ, and the point on circle, P(x,y), is found using polar coordinates,

$$x = R\cos(t)$$
$$y = R\sin(t)$$

Where R=$|\vec{r}|$ is the radius of the circle.

For example a circle with a radius of 5 is,

$$\vec{r}(t) = \langle 5\cos(t), 5\sin(t) \rangle$$

Circles in 3-D

In 3-D, circles contained in a coordinate plane can be described in a similar fashion as example #3. A circle of radius 3 in the x z-plane (y=0) can be written as:

$$\langle 3\cos(t), 0, 3\sin(t) \rangle$$

Additionally the center point can be moved. If this circle has a center point (2,6,8) and is parallel to the xz-plane.

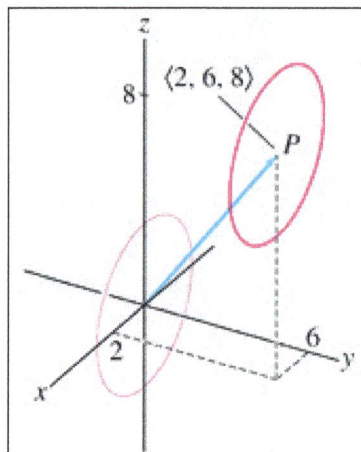

Then the vector form can be found using vector addition:

$$\vec{r}(t) = \langle 2,6,8 \rangle + \langle 3\cos(t), 0, 3\sin(t) \rangle$$
$$= \langle 2 + 3\cos(t), 6, 8 + 3\sin(t) \rangle$$

And the parametric equations for this circle are:

$$x = 2 + 3\cos(t)$$
$$y = 6$$
$$z = 8 + 3\sin(t)$$

Spiral Helix

A spiral helix with a radius of 1, circles in the xy-plane while it moves with a constant speed up the z-axis.

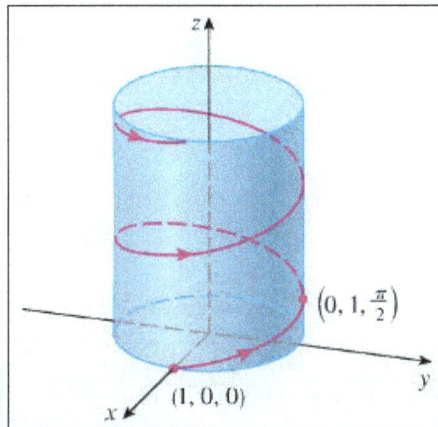

A parameterization of this path is,

$$\vec{r}(t) = \langle \cos(t), \sin(t), t \rangle$$

Ellipses

Ellipses are similar to circles, with the small change that the radius in the xx and yy direction may not be the same,

$$\vec{r}(t) = \langle A\cos(t), B\sin(t) \rangle$$

Where A and B are the semi-major and semi-minor axes.

A parameterization of an ellipse in the yz-plane with a semi-major axes of 6 along the z-axis and a semi-minor axis of 4 along the y-axis center at the point (−5,0,7) is:

$$\vec{r}(t) = \langle -5,0,7 \rangle + \langle 0, 4\cos(t), 6\sin(t) \rangle$$
$$= \langle -5, 4\cos(t), 7 + 6\sin(t) \rangle$$

Coordinate Plane Projections

Projections of a 3-D curve onto one of the coordinate planes, xy-pane, xz-plane or yz-plane, is one way to visualize what the graph looks like by looking at different 2-D graphs.

For example the helix:

$$\vec{r}(t) = \langle \sin(t), \cos(t), t \rangle$$

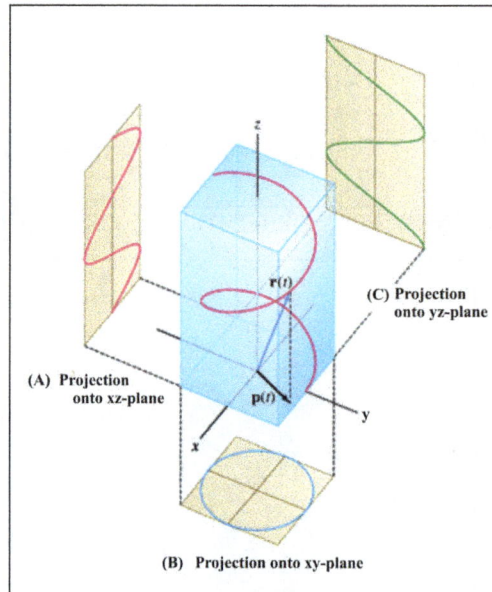

(A) Projection onto xz-plane

(B) Projection onto xy-plane

(C) Projection onto yz-plane

- The projection on the x y-plane (z = 0) is a circle,

  $$\langle -\sin(t), \cos(t), 0 \rangle$$

- The projection on the x z-plane (y = 0) is a sine wave,

  $$\langle -\sin(t), 0, t \rangle$$

- The projection on the y z-plane (x = 0) is a cosine wave,

  $$\langle 0, \cos(t), t \rangle.$$

Fundamental Theorems of Calculus

Some of the basic theorems in calculus are mean value theorem, Rolle's Theorem, extreme value theorem, Taylor's theorem and divergence theorem. Mean value theorem is used for proving statements regarding a function on an interval starting from local hypotheses about derivatives at points of the interval. This chapter has been carefully written to provide an easy understanding of the varied facets of these theories.

The Fundamental theorem of calculus links two branches; differential calculus (concerning rates of change and slopes of curves) and integral calculus (concerning the accumulation of quantities and the areas under and between curves).

First Fundamental Theorem of Calculus

If 'f' is a continuous function on the closed interval [a, b] and A (x) is the area function. Then $A'(x) = f(x)$, for all $x \in [a, b]$.

Second Fundamental Theorem of Calculus

If 'f' is a continuous function defined on the closed interval [a, b] and F is an anti-derivative of 'f'. Then,

$$\int_a^b f(x)\, dx = [F(x)]_b^a = F(b) - F(a)$$

Important Points to Remember:

- In words, the Theorem 2 tells us that $\int_a^b f(x)\, dx$ = (value of the anti-derivative 'F' of 'f' at the upper limit b) – (value of the same anti-derivative at the lower limit a).

- This theorem is useful because we can calculate the definite integral without calculating the limit of a sum.

- The most important step in evaluating a definite integral is finding a function whose derivative is equal to the integrand. It strengthens the relationship between differentiation and integration.

- In $\int_a^b f(x)\, dx$, the function 'f' should be well defined and continuous in [a, b].

Calculation Steps

Two simple steps can help you calculate $\int_a^b f(x)\, dx$ as shown below:

1. Find the indefinite integral $\int f(x)\, dx$. Let this be F(x). There is no need to keep the integration constant C because it disappears while evaluating the value of the definite integral.

2.   Evaluate F(b) − F(a) = [F (x)]$_a^b$. This is the value of $\int_a^b$ f(x) dx.

Let's look at some examples now.

Example:

Evaluate $\int_2^3$ x² dx

Solution: Let, I = $\int_2^3$ x² dx.

Now, the indefinite integral, $\int$ x² dx = x³/3 = F(x)

Using the second fundamental theorem of calculus, we get:

I = F(a) − F(b) = (3³/3) − (2³/3) = 27/3 − 8/3 = 19/3

Therefore, $\int_2^3$ x² dx = 19/3.

Example:

Evaluate $\int_4^9$ [√x / (30 − x$^{3/2}$)²] dx

Solution: Let I = $\int_4^9$ [√x / (30 − x$^{3/2}$)²] dx

First, we find the anti-derivative of the integrand.

Let's substitute (30 − x$^{3/2}$) = t, so that − 3/2 √x dx = dt or √x dx = − 2/3 dt.

Therefore, the indefinite integral,

$\int$ [√x / (30 − x$^{3/2}$)²] dx = − 2/3 $\int$ dt/t² = 2/3 (1/t)

= 2/3 [1 / (30 − x$^{3/2}$)] = F(x).

Hence, using the second fundamental theorem of calculus, we get:

I = F(9) − F(4) = 2/3 [1 / (30 − 27) − 1 / (30 − 8)] = 2/3 [1/3 − 1/22] = 19/99

Therefore,

$\int_4^9$ [√x / (30 − x$^{3/2}$)²] dx = 19/99

Example:

Evaluate $\int_1^2$ xdx / (x + 1)(x + 2)

Solution: Let I = $\int_1^2$ x dx / (x + 1)(x + 2)

Using the rules of partial fractions, we get:

x / (x + 1)(x + 2) = − 1 / (x + 1) + 2 / (x + 2)

Therefore, the indefinite integral,

$\int$ x dx / (x + 1)(x + 2) = − log |x + 1| + 2log |x + 2| = F(x)

Hence, using the second fundamental theorem of calculus, we get

$$I = F(2) - F(1) = [- \log 3 + 2\log 4] - [- \log 2 + 2\log 3]$$

$$= - 3\log 3 + \log 2 + 2\log 4 = \log (32/27)$$

Therefore,

$$\int_1^2 x dx / (x + 1)(x + 2) = \log (32/27)$$

Example:

Evaluate $\int_0^{\pi/4} \sin^3 2t \cos 2t \, dt$

Solution: Let $I = \int_0^{\pi/4} \sin^3 2t \cos 2t \, dt$

To solve the indefinite integral, let's substitute $\sin 2t = u$, so that $2 \cos 2t \, dt = du$ or $\cos 2t \, dt = \frac{1}{2} du$. Therefore,

$$\int \sin^3 2t \cos 2t \, dt = \frac{1}{2} \int u^3 \, du = (1/8) u^4 = (1/8) \sin^4 2t = F(x).$$

Hence, using the second fundamental theorem of calculus, we get:

$$I = F(\pi/4) - F(0) = 1/8 [\sin^4 (\pi/2) - \sin^4 0] = 1/8(1 - 0) = 1/8$$

Therefore, $\int_0^{\pi/4} \sin^3 2t \cos 2t \, dt = 1/8$.

MEAN VALUE THEOREM

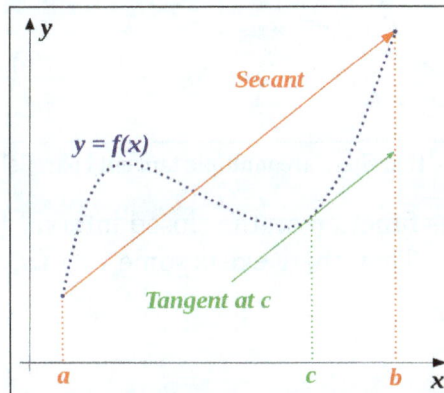

For any function that is continuous on $[a, b]$ and differentiable on (a, b) there exists some c in the interval (a, b) such that the secant joining the endpoints of the interval $[a, b]$ is parallel to the tangent at c.

In mathematics, the mean value theorem states, roughly, that for a given planar arc between two endpoints, there is at least one point at which the tangent to the arc is parallel to the secant through its endpoints.

This theorem is used to prove statements about a function on an interval starting from local hypotheses about derivatives at points of the interval.

More precisely, if f is a continuous function on the closed interval $[a,b]$, and differentiable on the open interval (a,b), then there exists a point c in (a,b) such that:

$$f'(c) = \frac{f(b)-f(a)}{b-a}.$$

It is one of the most important results in real analysis.

Formal Statement

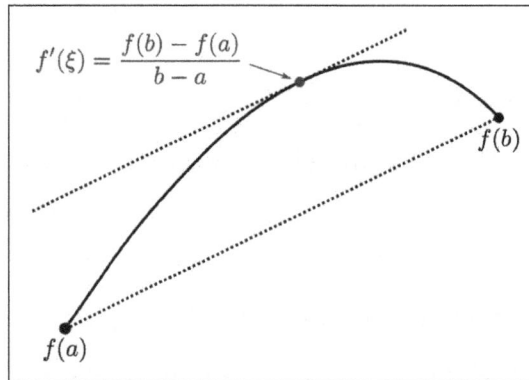

The function f attains the slope of the secant between a and b as the derivative at the point $\xi \in (a,b)$.

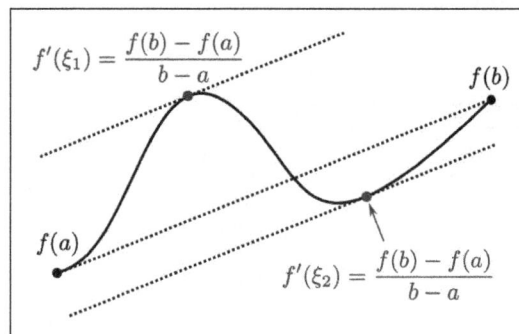

It is also possible that there are multiple tangents parallel to the secant.

Let $f:[a,b] \to \mathbb{R}$ be a continuous function on the closed interval $[a,b]$, and differentiable on the open interval (a,b), where $a < b$. Then there exists some c in (a,b) such that,

$$f'(c) = \frac{f(b)-f(a)}{b-a}.$$

The mean value theorem is a generalization of Rolle's theorem, which assumes $f(a) = f(b)$, so that the right-hand side above is zero.

The mean value theorem is still valid in a slightly more general setting. One only needs to assume that $f:[a,b] \to \mathbb{R}$ is continuous on $[a,b]$, and that for every x in (a,b) the limit,

$$\lim_{h \to 0} \frac{f(x+h)-f(x)}{h}$$

exists as a finite number or equals ∞ or $-\infty$. If finite, that limit equals $f'(x)$. An example where this version of the theorem applies is given by the real-valued cube root function mapping $x \to x^{\frac{1}{3}}$, whose derivative tends to infinity at the origin.

Note that the theorem, as stated, is false if a differentiable function is complex-valued instead of real-valued. For example, define $f(x) = e^{xi}$ for all real x. Then,

$$f(2\pi) - f(0) = 0 = 0(2\pi - 0)$$

while $f'(x) \neq 0$ for any real x.

These formal statements are also known as Lagrange's Mean Value Theorem.

Proof

The expression $\dfrac{f(b) - f(a)}{b - a}$ gives the slope of the line joining the points $(a, f(a))$ and $(b, f(b))$, which is a chord of the graph of f, while $f'(x)$ gives the slope of the tangent to the curve at the point $(x, f(x))$. Thus the mean value theorem says that given any chord of a smooth curve, we can find a point lying between the end-points of the chord such that the tangent at that point is parallel to the chord. The following proof illustrates this idea.

Define $g(x) = f(x) - rx$, where r is a constant. Since f is continuous on $[a, b]$ and differentiable on (a, b), the same is true for g. We now want to choose r so that g satisfies the conditions of Rolle's theorem. Namely,

$$g(a) = g(b) \Leftrightarrow f(a) - ra = f(b) - rb$$
$$\Leftrightarrow r(b - a) = f(b) - f(a)$$
$$\Leftrightarrow r = \frac{f(b) - f(a)}{b - a}.$$

By Rolle's theorem, since g is differentiable and $g(a) = g(b)$, there is some c in (a, b) for which $g'(c) = 0$, and it follows from the equality $g(x) = f(x) - rx$ that,

$$g'(x) = f'(x) - r$$
$$g'(c) = 0$$
$$g'(c) = f'(c) - r = 0$$
$$\Rightarrow f'(c) = r = \frac{f(b) - f(a)}{b - a}$$

A Simple Application

Assume that f is a continuous, real-valued function, defined on an arbitrary interval I of the real line. If the derivative of f at every interior point of the interval I exists and is zero, then f is constant in the interior.

Proof: Assume the derivative of f at every interior point of the interval I exists and is zero. Let (a, b) be an arbitrary open interval in I. By the mean value theorem, there exists a point c in (a,b) such that,

$$0 = f'(c) = \frac{f(b) - f(a)}{b - a}.$$

This implies that $f(a) = f(b)$. Thus, f is constant on the interior of I and thus is constant on I by continuity.

Remarks:

- Only continuity of f, not differentiability, is needed at the endpoints of the interval I. No hypothesis of continuity needs to be stated if I is an open interval, since the existence of a derivative at a point implies the continuity at this point.

- The differentiability of f can be relaxed to one-sided differentiability, Cauchy's mean value theorem.

Cauchy's mean value theorem, also known as the extended mean value theorem, is a generalization of the mean value theorem. It states: If functions f and g are both continuous on the closed interval $[a, b]$, and differentiable on the open interval (a, b), then there exists some $c \in (a, b)$ such that,

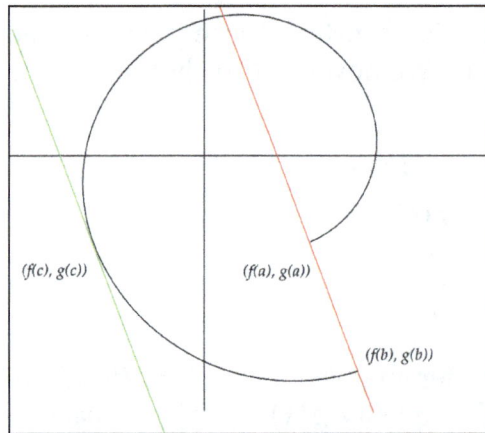

Geometrical meaning of Cauchy's theorem.

$$(f(b) - f(a))g'(c) = (g(b) - g(a))f'(c).$$

Of course, if $g(a) \neq g(b)$ and if $g'(c) \neq 0$, this is equivalent to:

$$\frac{f'(c)}{g'(c)} = \frac{f(b) - f(a)}{g(b) - g(a)}.$$

Geometrically, this means that there is some tangent to the graph of the curve,

$$\begin{cases} [a,b] \to \mathrm{R}^2 \\ t \mapsto (f(t), g(t)) \end{cases}$$

which is parallel to the line defined by the points ($f(a)$, $g(a)$) and ($f(b)$, $g(b)$). However Cauchy's theorem does not claim the existence of such a tangent in all cases where ($f(a)$, $g(a)$) and ($f(b)$, $g(b)$) are distinct points, since it might be satisfied only for some value c with $f'(c) = g'(c) = 0$, in other words a value for which the mentioned curve is stationary; in such points no tangent to the curve is likely to be defined at all. An example of this situation is the curve given by,

$$t \mapsto (t^3, 1-t^2),$$

which on the interval $[-1, 1]$ goes from the point $(-1, 0)$ to $(1, 0)$, yet never has a horizontal tangent; however it has a stationary point (in fact a cusp) at $t = 0$.

Cauchy's mean value theorem can be used to prove l'Hôpital's rule. The mean value theorem is the special case of Cauchy's mean value theorem when $g(t) = t$.

Proof of Cauchy's Mean Value Theorem

The proof of Cauchy's mean value theorem is based on the same idea as the proof of the mean value theorem.

- Suppose $g(a) \neq g(b)$. Define $h(x) = f(x) - rg(x)$, where r is fixed in such a way that $h(a) = h(b)$, namely

$$h(a) = h(b) \Leftrightarrow f(a) - rg(a) = f(b) - rg(b)$$
$$\Leftrightarrow r(g(b) - g(a)) = f(b) - f(a)$$
$$\Leftrightarrow r = \frac{f(b) - f(a)}{g(b) - g(a)}.$$

  Since f and g are continuous on $[a, b]$ and differentiable on (a, b), the same is true for h. All in all, h satisfies the conditions of Rolle's theorem: consequently, there is some c in (a, b) for which $h'(c) = 0$. Now using the definition of h we have:

$$0 = h'(c) = f'(c) - rg'(c) = f'(c) - \left(\frac{f(b) - f(a)}{g(b) - g(a)} \right) g'(c).$$

  Therefore:

$$f'(c) = \frac{f(b) - f(a)}{g(b) - g(a)} g'(c),$$

  which implies the result.

- If $g(a) = g(b)$, then, applying Rolle's theorem to g, it follows that there exists c in (a, b) for which $g'(c) = 0$. Using this choice of c, Cauchy's mean value theorem (trivially) holds.

Generalization for Determinants

Assume that f, g, and h are differentiable functions on (a,b) that are continuous on $[a,b]$.

Define:

$$D(x) = \begin{vmatrix} f(x) & g(x) & h(x) \\ f(a) & g(a) & h(a) \\ f(b) & g(b) & h(b) \end{vmatrix}$$

There exists $c \in (a,b)$ such that $D'(c) = 0$.

that,

$$D'(x) = \begin{vmatrix} f'(x) & g'(x) & h'(x) \\ f(a) & g(a) & h(a) \\ f(b) & g(b) & h(b) \end{vmatrix}$$

and if we place $h(x) = 1$, we get Cauchy's mean value theorem. If we place $h(x) = 1$ and $g(x)$ x we get Lagrange's mean value theorem.

The proof of the generalization is quite simple: each of $D(a)$ and $D(b)$ are determinants with two identical rows, hence $D(a) = D(b) = 0$. The Rolle's theorem implies that there exists $c \in (a,b)$ such that $D'(c) = 0$.

Mean Value Theorem in Several Variables

The mean value theorem generalizes to real functions of multiple variables. The trick is to use parametrization to create a real function of one variable, and then apply the one-variable theorem.

Let G be an open convex subset of $\mathbb{R}^n$, and let $f : G \rightarrow \mathbb{R}$ be a differentiable function. Fix points $x, y \in G$, and define $g(t) = f\big((1-t)x + ty\big)$. Since g is a differentiable function in one variable, the mean value theorem gives:

$$g(1) - g(0) = g'(c)$$

for some c between 0 and 1. But since $g(1) = f(y)$ and $g(0) = f(x)$, computing $g'(c)$ explicitly we have:

$$f(y) - f(x) = \nabla f\big((1-c)x + cy\big) \cdot (y - x)$$

where ∇ denotes a gradient and $\cdot$ a dot product. Note that this is an exact analog of the theorem in one variable (in the case $n = 1$ this *is* the theorem in one variable). By the Cauchy–Schwarz inequality, the equation gives the estimate:

$$\big| f(y) - f(x) \big| \leq \big| \nabla f\big((1-c)x + cy\big) \big| \big| y - x \big|.$$

In particular, when the partial derivatives of f are bounded, f is Lipschitz continuous (and therefore uniformly continuous). Note that f is not assumed to be continuously differentiable or continuous on the closure of G. However, in order to use the chain rule to compute g', we really do need to know that f is differentiable on G; the existence of the x and y partial derivatives by itself is not sufficient for the theorem to be true.

As an application of the above, we prove that f is constant if G is open and connected and every partial derivative of f is 0. Pick some point $x_0 \in G$, and let $g(x) = f(x) - f(x_0)$. We want to show $g(x) = 0$ for every $x \in G$. For that, let $E = \{x \in G : g(x) = 0\}$. Then E is closed and nonempty. It is open too: for every $x \in E$,

$$\left| g(y) \right| = \left| g(y) - g(x) \right| \leq (0) \left| y - x \right| = 0$$

for every y in some neighborhood of x. (Here, it is crucial that x and y are sufficiently close to each other.) Since G is connected, we conclude $E = G$.

The above arguments are made in a coordinate-free manner; hence, they generalize to the case when G is a subset of a Banach space.

Mean Value Theorem for Vector-valued Functions

There is no exact analog of the mean value theorem for vector-valued functions.

In *Principles of Mathematical Analysis,* Rudin gives an inequality which can be applied to many of the same situations to which the mean value theorem is applicable in the one dimensional case:

Theorem: *For a continuous vector-valued function* $f : [a,b] \to \mathbb{R}^k$ *differentiable on* (a,b), *there exists* $x \in (a,b)$ *such that* $| f'(x) | \geq \dfrac{1}{b-a} | f(b) - f(a) |$.,

Jean Dieudonné in his classic treatise *Foundations of Modern Analysis* discards the mean value theorem and replaces it by mean inequality as the proof is not constructive and one cannot find the mean value and in applications one only needs mean inequality. Serge Lang in *Analysis I* uses the mean value theorem, in integral form, as an instant reflex but this use requires the continuity of the derivative. If one uses the Henstock–Kurzweil integral one can have the mean value theorem in integral form without the additional assumption that derivative should be continuous as every derivative is Henstock–Kurzweil integrable. The problem is roughly speaking the following: If $f : U \to \mathbb{R}^m$ is a differentiable function (where $U \subset \mathbb{R}^n$ is open) and if $x + th, x, h \in \mathbb{R}^n, t \in [0, 1]$ is the line segment in question (lying inside U), then one can apply the above parametrization procedure to each of the component functions f_i ($i = 1, ..., m$) of f (in the above notation set $y = x + h$). In doing so one finds points $x + t_i h$ on the line segment satisfying,

$$f_i(x+h) - f_i(x) = \nabla f_i(x + t_i h) \cdot h.$$

But generally there will not be a *single* point $x + t^* h$ on the line segment satisfying,

$$f_i(x+h) - f_i(x) = \nabla f_i(x + t^* h) \cdot h.$$

for all i *simultaneously*. For example, define:

$$\begin{cases} f : [0, 2\pi] \to \mathbb{R}^2 \\ f(x) = (\cos(x), \sin(x)) \end{cases}$$

Then $f(2\pi) - f(0) = 0 \in \mathbb{R}^2$, but $f_1'(x) = -\sin(x)$ and $f_2'(x) = \cos(x)$ are never simultaneously zero as x ranges over $[0, 2\pi]$.

However a certain type of generalization of the mean value theorem to vector-valued functions is obtained as follows: Let f be a continuously differentiable real-valued function defined on an open interval I, and let x as well as $x + h$ be points of I. The mean value theorem in one variable tells us that there exists some t^* between 0 and 1 such that,

$$f(x+h) - f(x) = f'(x+t^*h) \cdot h.$$

On the other hand, we have, by the fundamental theorem of calculus followed by a change of variables,

$$f(x+h) - f(x) = \int_x^{x+h} f'(u)du = \left(\int_0^1 f'(x+th)dt \right) \cdot h.$$

Thus, the value $f'(x + t^*h)$ at the particular point t^* has been replaced by the mean value,

$$\int_0^1 f'(x+th)dt.$$

This last version can be generalized to vector valued functions:

Lemma: Let $U \subset \mathbb{R}^n$ be open, $f : U \to \mathbb{R}^m$ continuously differentiable, and $x \in U$, $h \in \mathbb{R}^n$ vectors such that the line segment $x + th$, $0 \le t \le 1$ remains in U. Then we have:

$$f(x+h) - f(x) = \left(\int_0^1 Df(x+th)dt \right) \cdot h,$$

where Df denotes the Jacobian matrix of f and the integral of a matrix is to be understood componentwise.

Proof: Let $f_1, ..., f_m$ denote the components of f and define:

$$\begin{cases} g_i : [0,1] \to \mathbb{R} \\ g_i(t) = f_i(x+th) \end{cases}$$

Then we have,

$$f_i(x+h) - f_i(x) = g_i(1) - g_i(0) = \int_0^1 g_{i'}(t)dt$$

$$= \int_0^1 \left(\sum_{j=1}^n \frac{\partial f_i}{\partial x_j}(x+th)h_j \right) dt = \sum_{j=1}^n \left(\int_0^1 \frac{\partial f_i}{\partial x_j}(x+th)dt \right) h_j.$$

The claim follows since Df is the matrix consisting of the components $\dfrac{\partial f_i}{\partial x_j}$.

Lemma: Let $v : [a, b] \to \mathbb{R}^m$ be a continuous function defined on the interval $[a, b] \subset \mathbb{R}$. Then we have:

$$\left\| \int_a^b v(t)dt \right\| \le \int_a^b \| v(t) \| dt.$$

Proof: Let u in $\mathbb{R}^m$ denote the value of the integral,

$$u := \int_a^b v(t)dt.$$

Now we have (using the Cauchy–Schwarz inequality):

$$\| u \|^2 = \langle u, u \rangle = \left\langle \int_a^b v(t)dt, u \right\rangle = \int_a^b \langle v(t), u \rangle dt \leqslant \int_a^b \| v(t) \| \cdot \| u \| dt = \| u \| \int_a^b \| v(t) \| dt$$

Now cancelling the norm of u from both ends gives us the desired inequality.

Mean Value Inequality. If the norm of $Df(x + th)$ is bounded by some constant M for t in $[0, 1]$, then:

$$\| f(x + h) - f(x) \| \leqslant M \| h \|.$$

Proof: From Lemma 1 and 2 it follows that,

$$\| f(x + h) - f(x) \| = \left\| \int_0^1 (Df(x + th) \cdot h)dt \right\| \leqslant \int_0^1 \| Df(x + th) \| \cdot \| h \| dt \leqslant M \| h \|.$$

Mean Value Theorems for Definite Integrals

First Mean Value Theorem for Definite Integrals

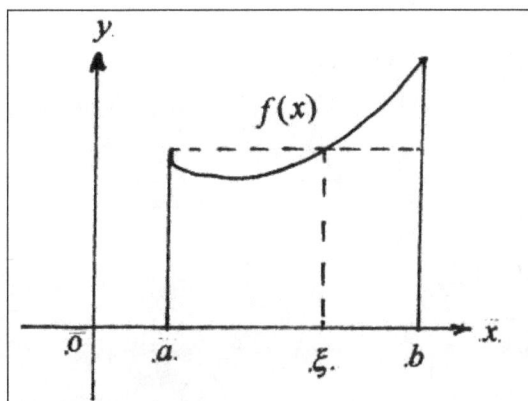

Geometrically: interpreting f(c) as the height of a rectangle and b–a as the width, this rectangle has the same area as the region below the curve from a to b

Let $f: [a, b] \to \mathbb{R}$ be a continuous function. Then there exists c in (a, b) such that,

$$\int_a^b f(x)dx = f(c)(b - a).$$

Since the mean value of f on $[a, b]$ is defined as,

$$\frac{1}{b-a} \int_a^b f(x)dx,$$

we can interpret the conclusion as f achieves its mean value at some c in (a, b).

In general, if $f : [a, b] \to \mathrm{R}$ is continuous and g is an integrable function that does not change sign on $[a, b]$, then there exists c in (a, b) such that,

$$\int_a^b f(x)g(x)dx = f(c)\int_a^b g(x)dx.$$

Proof of the First Mean Value Theorem for Definite Integrals

Suppose $f : [a, b] \to \mathrm{R}$ is continuous and g is a nonnegative integrable function on $[a, b]$. By the extreme value theorem, there exists m and M such that for each x in $[a, b]$, $m \leqslant f(x) \leqslant M$ and $f[a, b] = [m, M]$. Since g is nonnegative,

$$m\int_a^b g(x)dx \leqslant \int_a^b f(x)g(x)dx \leqslant M\int_a^b g(x)dx.$$

Now let,

$$I = \int_a^b g(x)dx.$$

If $I = 0$, we're done since,

$$0 \leqslant \int_a^b f(x)g(x)dx \leqslant 0$$

means,

$$\int_a^b f(x)g(x)dx = 0,$$

so for any c in (a, b),

$$\int_a^b f(x)g(x)dx = f(c)I = 0.$$

If $I \neq 0$, then:

$$m \leqslant \frac{1}{I}\int_a^b f(x)g(x)dx \leqslant M.$$

By the intermediate value theorem, f attains every value of the interval $[m, M]$, so for some c in $[a, b]$,

$$f(c) = \frac{1}{I}\int_a^b f(x)g(x)dx,$$

that is,

$$\int_a^b f(x)g(x)dx = f(c)\int_a^b g(x)dx.$$

Finally, if g is negative on $[a, b]$, then,

$$M \int_a^b g(x)dx \leqslant \int_a^b f(x)g(x)dx \leqslant m \int_a^b g(x)dx$$

and we still get the same result as above.

Second mean Value Theorem for Definite Integrals

There are various slightly different theorems called the second mean value theorem for definite integrals. A commonly found version is as follows:

If $G : [a, b] \to R$ is a positive monotonically decreasing function and $\varphi : [a, b] \to R$ is an integrable function, then there exists a number x in $[a, b]$ such that,

$$\int_a^b G(t)\varphi(t)dt = G(a^+) \int_a^x \varphi(t)dt.$$

Here $G(a^+)$ stands for $\lim\limits_{x \to a^+} G(x)$, the existence of which follows from the conditions. Note that it is essential that the interval $(a, b]$ contains b. A variant not having this requirement is:

If $G : [a, b] \to R$ is a monotonic (not necessarily decreasing and positive) function and $\varphi : [a, b] \to R$ is an integrable function, then there exists a number x in (a, b) such that,

$$\int_a^b G(t)\varphi(t)dt = G(a^+) \int_a^x \varphi(t)dt + G(b^-) \int_x^b \varphi(t)dt.$$

Mean Value Theorem for Integration Fails for Vector-valued Functions

If the function G returns a multi-dimensional vector, then the MVT for integration is not true, even if the domain of G is also multi-dimensional.

For example, consider the following 2-dimensional function defined on an n-dimensional cube:

$$\begin{cases} G : [0, 2\pi]^n \to \mathbb{R}^2 \\ G(x_1, \cdots, x_n) = \left(\sin(x_1 + \cdots + x_n), \cos(x_1 + \cdots + x_n) \right) \end{cases}$$

Then, by symmetry it is easy to see that the mean value of G over its domain is (0,0):

$$\int_{[0, 2\pi]^n} G(x_1, \cdots, x_n)dx_1 \cdots dx_n = (0, 0)$$

However, there is no point in which $G = (0, 0)$, because $|G| = 1$ everywhere.

A Probabilistic Analogue of the Mean Value Theorem

Let X and Y be non-negative random variables such that $E[X] < E[Y] < \infty$ and $X \leqslant_{st} Y$ (i.e. X is smaller than Y in the usual stochastic order). Then there exists an absolutely continuous non-negative random variable Z having probability density function:

$$f_Z(x) = \frac{\Pr(Y > x) - \Pr(X > x)}{E[Y] - E[X]}, \qquad x \geqslant 0.$$

Let g be a measurable and differentiable function such that $E[g(X)]$, $E[g(Y)] < \infty$, and let its derivative g' be measurable and Riemann-integrable on the interval $[x, y]$ for all $y \geq x \geq 0$. Then, $E[g'(Z)]$ is finite and,

$$E[g(Y)] - E[g(X)] = E[g'(Z)][E(Y) - E(X)].$$

Generalization in Complex Analysis

As noted above, the theorem does not hold for differentiable complex-valued functions. Instead, a generalization of the theorem is stated such:

Let $f : \Omega \to C$ be a holomorphic function on the open convex set Ω, and let a and b be distinct points in Ω. Then there exist points u, v on L_{ab} (the line segment from a to b) such that:

$$\mathrm{Re}(f'(u)) = \mathrm{Re}\left(\frac{f(b) - f(a)}{b - a} \right),$$

$$\mathrm{Im}(f'(v)) = \mathrm{Im}\left(\frac{f(b) - f(a)}{b - a} \right).$$

Where Re() is the Real part and Im() is the Imaginary part of a complex-valued function.

ROLLE'S THEOREM

In calculus, Rolle's theorem or Rolle's lemma essentially states that any real-valued differentiable function that attains equal values at two distinct points must have at least one stationary point somewhere between them—that is, a point where the first derivative (the slope of the tangent line to the graph of the function) is zero.

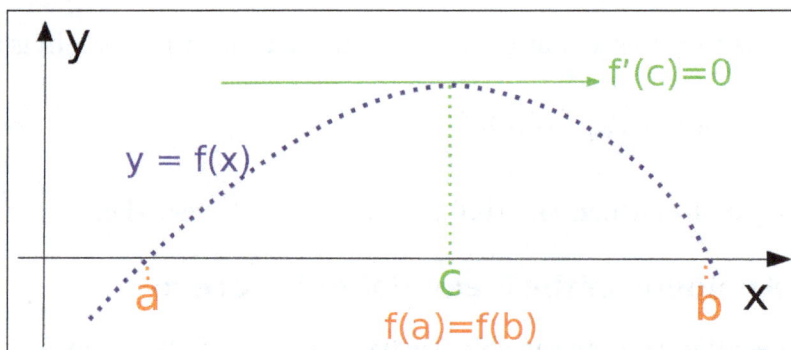

If a real-valued function f is continuous on a closed interval $[a, b]$, differentiable on the open interval (a, b), and $f(a) = f(b)$, then there exists a c in the open interval (a, b) such that $f'(c) = 0$.

Standard Version of the Theorem

If a real-valued function f is continuous on a proper closed interval $[a, b]$, differentiable on the open interval (a, b), and $f(a) = f(b)$, then there exists at least one c in the open interval (a, b) such that,

$$f'(c) = 0.$$

This version of Rolle's theorem is used to prove the mean value theorem, of which Rolle's theorem is indeed a special case. It is also the basis for the proof of Taylor's theorem.

Example:

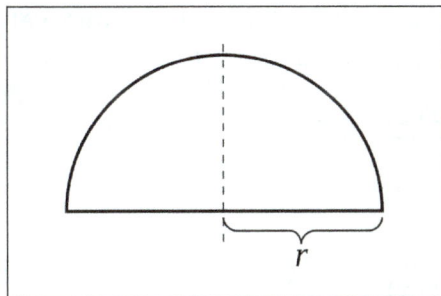

A semicircle of radius r.

For a radius $r > 0$, consider the function,

$$f(x) = \sqrt{r^2 - x^2}, \quad x \in [-r, r]$$

Its graph is the upper semicircle centered at the origin. This function is continuous on the closed interval $[-r, r]$ and differentiable in the open interval $(-r, r)$, but not differentiable at the endpoints $-r$ and r. Since $f(-r) = f(r)$, Rolle's theorem applies, and indeed, there is a point where the derivative of f is zero. Note that the theorem applies even when the function cannot be differentiated at the endpoints because it only requires the function to be differentiable in the open interval.

Example:

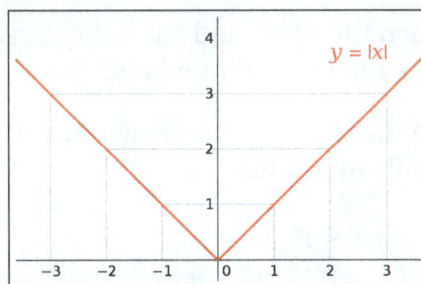

The graph of the absolute value function.

If differentiability fails at an interior point of the interval, the conclusion of Rolle's theorem may not hold. Consider the absolute value function:

$$f(x) = |x|, \quad x \in [-1, 1].$$

Then $f(-1) = f(1)$, but there is no c between -1 and 1 for which the $f'(c)$ is zero. This is because that function, although continuous, is not differentiable at $x = 0$. Note that the derivative of f changes its sign at $x = 0$, but without attaining the value 0. The theorem cannot be applied to this function because it does not satisfy the condition that the function must be differentiable for every x in the open interval. However, when the differentiability requirement is dropped from Rolle's theorem, f will still have a critical number in the open interval (a, b), but it may not yield a horizontal tangent (as in the case of the absolute value represented in the graph).

Generalization

The second example illustrates the following generalization of Rolle's theorem:

Consider a real-valued, continuous function f on a closed interval $[a, b]$ with $f(a) = f(b)$. If for every x in the open interval (a, b) the right-hand limit,

$$f'(x^+) := \lim_{h \to 0^+} \frac{f(x+h) - f(x)}{h}$$

and the left-hand limit,

$$f'(x^-) := \lim_{h \to 0^-} \frac{f(x+h) - f(x)}{h}$$

exist in the extended real line $[-\infty, \infty]$, then there is some number c in the open interval (a, b) such that one of the two limits,

$$f'(c^+) \text{ and } f'(c^-)$$

is ≥ 0 and the other one is ≤ 0 (in the extended real line). If the right- and left-hand limits agree for every x, then they agree in particular for c, hence the derivative of f exists at c and is equal to zero.

Remarks:

- If f is convex or concave, then the right- and left-hand derivatives exist at every inner point, hence the above limits exist and are real numbers.

- This generalized version of the theorem is sufficient to prove convexity when the one-sided derivatives are monotonically increasing:

$$f'(x^-) \leq f'(x^+) \leq f'(y^-), \qquad x < y$$

Proof of the Generalized Version

Since the proof for the standard version of Rolle's theorem and the generalization are very similar, we prove the generalization.

The idea of the proof is to argue that if $f(a) = f(b)$, then f must attain either a maximum or a minimum somewhere between a and b, say at c, and the function must change from increasing to decreasing (or the other way around) at c. In particular, if the derivative exists, it must be zero at c.

By assumption, f is continuous on $[a, b]$, and by the extreme value theorem attains both its maximum and its minimum in $[a, b]$. If these are both attained at the endpoints of $[a, b]$, then f is constant on $[a, b]$ and so the derivative of f is zero at every point in (a, b).

Suppose then that the maximum is obtained at an interior point c of (a, b) (the argument for the minimum is very similar, just consider $-f$). We shall examine the above right- and left-hand limits separately.

For a real h such that $c + h$ is in $[a, b]$, the value $f(c + h)$ is smaller or equal to $f(c)$ because f attains its maximum at c. Therefore, for every $h > 0$,

$$\frac{f(c+h)-f(c)}{h} \leq 0,$$

hence,

$$f'(c^+) := \lim_{h \to 0^+} \frac{f(c+h)-f(c)}{h} \leq 0,$$

where the limit exists by assumption, it may be minus infinity.

Similarly, for every $h < 0$, the inequality turns around because the denominator is now negative and we get,

$$\frac{f(c+h)-f(c)}{h} \geq 0,$$

hence,

$$f'(c^-) := \lim_{h \to 0^-} \frac{f(c+h)-f(c)}{h} \geq 0,$$

where the limit might be plus infinity.

Finally, when the above right- and left-hand limits agree (in particular when f is differentiable), then the derivative of f at c must be zero.

(Alternatively, we can apply Fermat's stationary point theorem directly.)

Generalization to Higher Derivatives

We can also generalize Rolle's theorem by requiring that f has more points with equal values and greater regularity. Specifically, suppose that:

- The function f is $n - 1$ times continuously differentiable on the closed interval $[a, b]$ and the nth derivative exists on the open interval (a, b), and

- There are n intervals given by $a_1 < b_1 \le a_2 < b_2 \le \ldots \le a_n < b_n$ in $[a, b]$ such that $f(a_k) = f(b_k)$ for every k from 1 to n. Then there is a number c in (a, b) such that the nth derivative of f at c is zero.

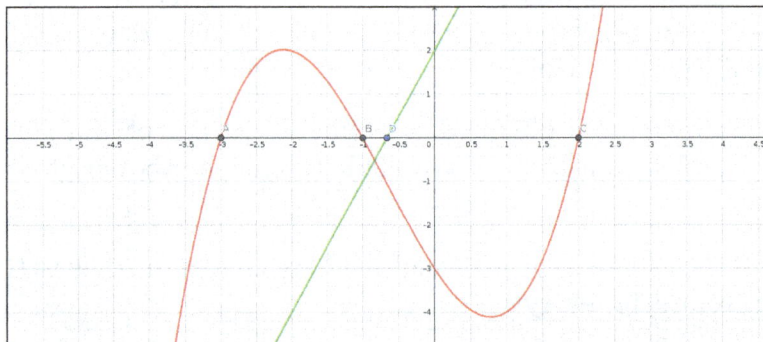

The red curve is the graph of function with 3 roots in the interval $[-3, 2]$. Thus its second derivative (graphed in green) also has a root in the same interval.

The requirements concerning the nth derivative of f can be weakened as in the generalization above, giving the corresponding (possibly weaker) assertions for the right- and left-hand limits defined above with $f^{(n-1)}$ in place of f.

Particularly, this version of the theorem asserts that if a function differentiable enough times has n roots (so they have the same value, that is 0), then there is an internal point where $f^{(n-1)}$ vanishes.

Proof

The proof uses mathematical induction. The case $n = 1$ is simply the standard version of Rolle's theorem. As the induction hypothesis, assume the generalization is true for $n - 1$. We want to prove it for $n > 1$. By the standard version of Rolle's theorem, for every integer k from 1 to n, there exists a c_k in the open interval (a_k, b_k) such that $f'(c_k) = 0$. Hence, the first derivative satisfies the assumptions on the $n - 1$ closed intervals $[c_1, c_2], \ldots, [c_{n-1}, c_n]$. By the induction hypothesis, there is a c such that the $(n - 1)$st derivative of f' at c is zero.

Generalizations to other Fields

Rolle's theorem is a property of differentiable functions over the real numbers, which are an ordered field. As such, it does not generalize to other fields, but the following corollary does: if a real polynomial factors (has all of its roots) over the real numbers, then its derivative does as well. One may call this property of a field Rolle's property. More general fields do not always have differentiable functions, but they do always have polynomials, which can be symbolically differentiated. Similarly, more general fields may not have an order, but one has a notion of a root of a polynomial lying in a field.

Thus Rolle's theorem shows that the real numbers have Rolle's property. Any algebraically closed field such as the complex numbers has Rolle's property. However, the rational numbers do not – for example, $x^3 - x = x(x - 1)(x + 1)$ factors over the rationals, but its derivative,

$$3x^2 - 1 = 3\left(x - \tfrac{1}{\sqrt{3}}\right)\left(x + \tfrac{1}{\sqrt{3}}\right)$$

does not. The question of which fields satisfy Rolle's property. For finite fields, the answer is that only F^2 and F^4 have Rolle's property.

L'HÔPITAL'S RULE

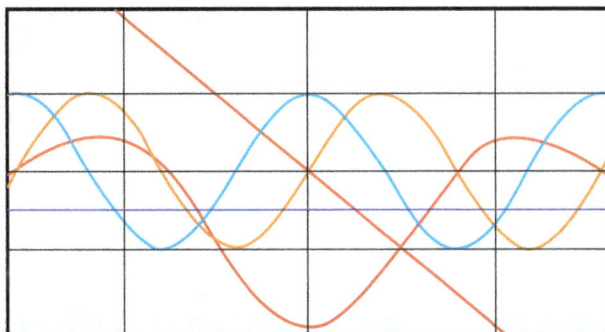

Example application of l'Hôpital's rule to $f(x) = \sin(x)$ and $g(x) = -0.5x$: the function $h(x) = f(x)/g(x)$ is undefined at $x = 0$, but can be completed to a continuous function on whole R by defining $h(0) = f'(0)/g'(0) = -2$.

In mathematics, and more specifically in calculus, L'Hôpital's rule or L'Hospital's rule (French: [lopital]) uses derivatives to help evaluate limits involving indeterminate forms. Application (or repeated application) of the rule often converts an indeterminate form to an expression that can be evaluated by substitution, allowing easier evaluation of the limit. The rule is named after the 17th-century French mathematician Guillaume de l'Hôpital. Although the contribution of the rule is often attributed to L'Hôpital, the theorem was first introduced to L'Hôpital in 1694 by the Swiss mathematician Johann Bernoulli.

L'Hôpital's rule states that for functions f and g which are differentiable on an open interval I except possibly at a point c contained in I, if $\lim_{x \to c} f(x) = \lim_{x \to c} g(x) = 0$ or $\pm\infty$, $g'(x) \neq 0$ for all x in I with $x \neq c$, and $\lim_{x \to c} \dfrac{f'(x)}{g'(x)}$ exists then,

$$\lim_{x \to c} \frac{f(x)}{g(x)} = \lim_{x \to c} \frac{f'(x)}{g'(x)}.$$

The differentiation of the numerator and denominator often simplifies the quotient or converts it to a limit that can be evaluated directly.

General Form

The general form of L'Hôpital's rule covers many cases. Let c and L be extended real numbers (i.e., real numbers, positive infinity, or negative infinity). Let I be an open interval containing c (for a two-sided limit) or an open interval with endpoint c (for a one-sided limit, or a limit at infinity if c is infinite). The real valued functions f and g are assumed to be differentiable on I except possibly at c, and additionally $g'(x) \neq 0$ on I except possibly at c. It is also assumed that $\lim_{x \to c} \dfrac{f'(x)}{g'(x)} = L$ Thus the rule applies to situations in which the ratio of the derivatives has a finite or infinite limit, but not to situations in which that ratio fluctuates permanently as x gets closer and closer to c.

If either,

$$\lim_{x \to c} f(x) = \lim_{x \to c} g(x) = 0$$

or

$$\lim_{x \to c} |f(x)| = \lim_{x \to c} |g(x)| = \infty,$$

then,

$$\lim_{x \to c} \frac{f(x)}{g(x)} = L.$$

Although we have written $x \to c$ throughout, the limits may also be one-sided limits ($x \to c^+$ or $x \to c^-$), when c is a finite endpoint of I.

In the second case, the hypothesis that f diverges to infinity is not used in the proof; thus, while the conditions of the rule are normally stated as above, the second sufficient condition for the rule's procedure to be valid can be more briefly stated as:

$$\lim_{x \to c} |g(x)| = \infty$$

The hypothesis that $g'(x) \neq 0$ appears most commonly in the literature, but some authors side-step this hypothesis by adding other hypotheses elsewhere. One method is to define the limit of a function with the additional requirement that the limiting function is defined everywhere on the relevant interval I except possibly at c. Another method is to require that both f and g be differentiable everywhere on an interval containing c.

Requirement that the Limit Exist

The requirement that the limit,

$$\lim_{x \to c} \frac{f'(x)}{g'(x)}$$

must exist is essential. Without this condition, f' or g' may exhibit undampened oscillations as x approaches c, in which case L'Hôpital's rule does not apply. For example, if $f(x) = x + \sin(x)$, $g(x) = x$ and $c = \pm\infty$, then,

$$\frac{f'(x)}{g'(x)} = \frac{1 + \cos(x)}{1}$$

this expression does not approach a limit as x goes to c, since the cosine function oscillates between 1 and −1. But working with the original functions, $\lim_{x \to \infty} \frac{f(x)}{g(x)}$.

Can be shown to exist:

$$\lim_{x\to\infty}\frac{f(x)}{g(x)} = \lim_{x\to\infty}\left(1+\frac{\sin(x)}{x}\right) = 1.$$

In a case such as this, all that can be concluded is that,

$$\liminf_{x\to c}\frac{f'(x)}{g'(x)} \le \liminf_{x\to c}\frac{f(x)}{g(x)} \le \limsup_{x\to c}\frac{f(x)}{g(x)} \le \limsup_{x\to c}\frac{f'(x)}{g'(x)},$$

so that if the limit of f/g exists, then it must lie between the inferior and superior limits of f'/g'. (In the example above, this is true, since 1 indeed lies between 0 and 2.)

Examples:

- Here is a basic example involving the exponential function, which involves the indeterminate form 0/0 at $x = 0$:

$$\lim_{x\to 0}\frac{e^x-1}{x^2+x} = \lim_{x\to 0}\frac{\dfrac{d}{dx}(e^x-1)}{\dfrac{d}{dx}(x^2+x)}$$

$$= \lim_{x\to 0}\frac{e^x}{2x+1}$$

$$= 1.$$

- This is a more elaborate example involving 0/0. Applying L'Hôpital's rule a single time still results in an indeterminate form. In this case, the limit may be evaluated by applying the rule three times:

$$\lim_{x\to 0}\frac{2\sin(x)-\sin(2x)}{x-\sin(x)} = \lim_{x\to 0}\frac{2\cos(x)-2\cos(2x)}{1-\cos(x)}$$

$$= \lim_{x\to 0}\frac{-2\sin(x)+4\sin(2x)}{\sin(x)}$$

$$= \lim_{x\to 0}\frac{-2\cos(x)+8\cos(2x)}{\cos(x)}$$

$$= \frac{-2+8}{1}$$

$$= 6.$$

- Here is an example involving ∞/∞:

$$\lim_{x\to\infty}x^n\cdot e^{-x} = \lim_{x\to\infty}\frac{x^n}{e^x} = \lim_{x\to\infty}\frac{nx^{n-1}}{e^x} = n\cdot\lim_{x\to\infty}\frac{x^{n-1}}{e^x}.$$

Repeatedly apply L'Hôpital's rule until the exponent is zero (if n is an integer) or negative (if n is fractional) to conclude that the limit is zero.

- Here is an example involving the indeterminate form $0 \cdot \infty$, which is rewritten as the form ∞/∞:

$$\lim_{x \to 0^+} x \ln x = \lim_{x \to 0^+} \frac{\ln x}{\dfrac{1}{x}} = \lim_{x \to 0^+} \frac{\dfrac{1}{x}}{-\dfrac{1}{x^2}} = \lim_{x \to 0^+} -x = 0.$$

- Here is an example involving the mortgage repayment formula and $0/0$. Let P be the principal (loan amount), r the interest rate per period and n the number of periods. When r is zero, the repayment amount per period is $\dfrac{P}{n}$ (since only principal is being repaid); this is consistent with the formula for non-zero interest rates:

$$\lim_{r \to 0} \frac{Pr(1+r)^n}{(1+r)^n - 1} = P \lim_{r \to 0} \frac{(1+r)^n + rn(1+r)^{n-1}}{n(1+r)^{n-1}}$$

$$= \frac{P}{n}.$$

- One can also use L'Hôpital's rule to prove the following theorem. If f is twice-differentiable in a neighborhood of x then,

$$\lim_{h \to 0} \frac{f(x+h) + f(x-h) - 2f(x)}{h^2} = \lim_{h \to 0} \frac{f'(x+h) - f'(x-h)}{2h}$$

$$= \lim_{h \to 0} \frac{f''(x+h) + f''(x-h)}{2}$$

$$= f''(x).$$

- Sometimes L'Hôpital's rule is invoked in a tricky way: suppose $f(x) + f'(x)$ converges as $x \to \infty$ and that $e^x \cdot f(x)$ converges to positive or negative infinity. Then:

$$\lim_{x \to \infty} f(x) = \lim_{x \to \infty} \frac{e^x \cdot f(x)}{e^x} = \lim_{x \to \infty} \frac{e^x \left(f(x) + f'(x) \right)}{e^x} = \lim_{x \to \infty} \left(f(x) + f'(x) \right)$$

and so, $\lim\limits_{x \to \infty} f(x)$ exists and $\lim\limits_{x \to \infty} f'(x) = 0$.

The result remains true without the added hypothesis that $e^x \cdot f(x)$ converges to positive or negative infinity, but the justification is then incomplete.

Complications

Sometimes L'Hôpital's rule does not lead to an answer in a finite number of steps unless some additional steps are applied. Examples include the following:

- Two applications can lead to a return to the original expression that was to be evaluated:

$$\lim_{x \to \infty} \frac{e^x + e^{-x}}{e^x - e^{-x}} = \lim_{x \to \infty} \frac{e^x - e^{-x}}{e^x + e^{-x}} = \lim_{x \to \infty} \frac{e^x + e^{-x}}{e^x - e^{-x}} = \cdots.$$

This situation can be dealt with by substituting $y = e^x$ and noting that y goes to infinity as x goes to infinity; with this substitution, this problem can be solved with a single application of the rule:

$$\lim_{x \to \infty} \frac{e^x + e^{-x}}{e^x - e^{-x}} = \lim_{y \to \infty} \frac{y + y^{-1}}{y - y^{-1}} = \lim_{y \to \infty} \frac{1 - y^{-2}}{1 + y^{-2}} = \frac{1}{1} = 1.$$

Alternatively, the numerator and denominator can both be multiplied by e^x at which point L'Hôpital's rule can immediately be applied successfully:

$$\lim_{x \to \infty} \frac{e^x + e^{-x}}{e^x - e^{-x}} = \lim_{x \to \infty} \frac{e^{2x} + 1}{e^{2x} - 1} = \lim_{x \to \infty} \frac{2e^{2x}}{2e^{2x}} = 1.$$

- An arbitrarily large number of applications may never lead to an answer even without repeating:

$$\lim_{x \to \infty} \frac{x^{\frac{1}{2}} + x^{-\frac{1}{2}}}{x^{\frac{1}{2}} - x^{-\frac{1}{2}}} = \lim_{x \to \infty} \frac{\frac{1}{2} x^{-\frac{1}{2}} - \frac{1}{2} x^{-\frac{3}{2}}}{\frac{1}{2} x^{-\frac{1}{2}} + \frac{1}{2} x^{-\frac{3}{2}}} = \lim_{x \to \infty} \frac{-\frac{1}{4} x^{-\frac{3}{2}} + \frac{3}{4} x^{-\frac{5}{2}}}{-\frac{1}{4} x^{-\frac{3}{2}} - \frac{3}{4} x^{-\frac{5}{2}}} = \cdots.$$

This situation too can be dealt with by a transformation of variables, in this case $y = \sqrt{x}$:

$$\lim_{x \to \infty} \frac{x^{\frac{1}{2}} + x^{-\frac{1}{2}}}{x^{\frac{1}{2}} - x^{-\frac{1}{2}}} = \lim_{y \to \infty} \frac{y + y^{-1}}{y - y^{-1}} = \lim_{y \to \infty} \frac{1 - y^{-2}}{1 + y^{-2}} = \frac{1}{1} = 1.$$

Again, an alternative approach is to multiply numerator and denominator by $x^{1/2}$ before applying L'Hôpital's rule:

$$\lim_{x \to \infty} \frac{x^{\frac{1}{2}} + x^{-\frac{1}{2}}}{x^{\frac{1}{2}} - x^{-\frac{1}{2}}} = \lim_{x \to \infty} \frac{x + 1}{x - 1} = \lim_{x \to \infty} \frac{1}{1} = 1.$$

A common pitfall is using L'Hôpital's rule with some circular reasoning to compute a derivative via a difference quotient. For example, consider the task of proving the derivative formula for powers of x:

$$\lim_{h \to 0} \frac{(x + h)^n - x^n}{h} = nx^{n-1}.$$

Applying L'Hôpital's rule and finding the derivatives with respect to h of the numerator and the denominator yields $n x^{n-1}$ as expected. However, differentiating the numerator required the use of the very fact that is being proven. This is an example of begging the question, since one may not assume the fact to be proven during the course of the proof.

Counter Examples when the Derivative of the Denominator is Zero

The necessity of the condition that $g'(x) \neq 0$ near c can be seen by the following counterexample due to Otto Stolz. Let $f(x) = x + \sin x \cos x$ and $g(x) = f(x)e^{\sin x}$ Then there is no limit for $f(x)/g(x)$ as $x \to \infty$ However,

$$\frac{f'(x)}{g'(x)} = \frac{2\cos^2 x}{(2\cos^2 x)e^{\sin x} + (x + \sin x \cos x)e^{\sin x}\cos x}$$

$$= \frac{2\cos x}{2\cos x + x + \sin x \cos x}e^{-\sin x},$$

which tends to 0 as $x \to \infty$.

Other Indeterminate Forms

Other indeterminate forms, such as 1^∞, 0^0, ∞^0, $0 \cdot \infty$, and $\infty - \infty$, can sometimes be evaluated using L'Hôpital's rule. For example, to evaluate a limit involving $\infty - \infty$, convert the difference of two functions to a quotient:

$$\lim_{x \to 1}\left(\frac{x}{x-1} - \frac{1}{\ln x}\right) = \lim_{x \to 1}\frac{x \cdot \ln x - x + 1}{(x-1) \cdot \ln x}$$

$$= \lim_{x \to 1}\frac{\ln x}{\dfrac{x-1}{x} + \ln x}$$

$$= \lim_{x \to 1}\frac{x \cdot \ln x}{x - 1 + x \cdot \ln x}$$

$$= \lim_{x \to 1}\frac{1 + \ln x}{1 + 1 + \ln x}$$

$$= \lim_{x \to 1}\frac{1 + \ln x}{2 + \ln x}$$

$$= \frac{1}{2},$$

where L'Hôpital's rule is applied when going from $\lim_{x \to 1}\left(\dfrac{x}{x-1} - \dfrac{1}{\ln x}\right) = \lim_{x \to 1}\dfrac{x \cdot \ln x - x + 1}{(x-1) \cdot \ln x}$ to $\lim_{x \to 1}\dfrac{\ln x}{\dfrac{x-1}{x} + \ln x}$

and again when going from $\lim_{x \to 1}\dfrac{x \cdot \ln x}{x - 1 + x \cdot \ln x}$ to $\lim_{x \to 1}\dfrac{1 + \ln x}{1 + 1 + \ln x}$.

L'Hôpital's rule can be used on indeterminate forms involving exponents by using logarithms to "move the exponent down". Here is an example involving the indeterminate form 0^0:

$$\lim_{x \to 0^+} x^x = \lim_{x \to 0^+} e^{\ln(x^x)} = \lim_{x \to 0^+} e^{x \cdot \ln x} = e^{\lim_{x \to 0^+}(x \cdot \ln x)}.$$

It is valid to move the limit inside the exponential function because the exponential function is continuous. Now the exponent x has been "moved down". The limit $\lim\limits_{x \to 0^+} x \cdot \ln x$ is of the indeterminate form $0 \cdot \infty$, but as shown in an example above, l'Hôpital's rule may be used to determine that

$$\lim_{x \to 0^+} x \cdot \ln x = 0.$$

Thus,

$$\lim_{x \to 0^+} x^x = e^0 = 1.$$

Geometric Interpretation

Consider the curve in the plane whose x-coordinate is given by $g(t)$ and whose y-coordinate is given by $f(t)$, with both functions continuous, i.e., the locus of points of the form $[g(t), f(t)]$. Suppose $f(c) = g(c) = 0$. The limit of the ratio $f(t)/g(t)$ as $t \to c$ is the slope of the tangent to the curve at the point $[g(c), f(c)] = [0,0]$. The tangent to the curve at the point $[g(t), f(t)]$ is given by $[g'(t), f'(t)]$. L'Hôpital's rule then states that the slope of the curve when $t = c$ is the limit of the slope of the tangent to the curve as the curve approaches the origin, provided that this is defined.

Proof of L'Hôpital's Rule

Special Case

The proof of L'Hôpital's rule is simple in the case where f and g are continuously differentiable at the point c and where a finite limit is found after the first round of differentiation. It is not a proof of the general L'Hôpital's rule because it is stricter in its definition, requiring both differentiability and that c be a real number. Since many common functions have continuous derivatives (e.g. polynomials, sine and cosine, exponential functions), it is a special case worthy of attention.

Suppose that f and g are continuously differentiable at a real number c, that $f(c) = g(c) = 0$, and that $g'(c) \neq 0$. Then,

$$\lim_{x \to c} \frac{f(x)}{g(x)} = \lim_{x \to c} \frac{f(x)-0}{g(x)-0} = \lim_{x \to c} \frac{f(x)-f(c)}{g(x)-g(c)}$$

$$= \lim_{x \to c} \frac{\left(\dfrac{f(x)-f(c)}{x-c} \right)}{\left(\dfrac{g(x)-g(c)}{x-c} \right)} = \frac{\lim\limits_{x \to c} \left(\dfrac{f(x)-f(c)}{x-c} \right)}{\lim\limits_{x \to c} \left(\dfrac{g(x)-g(c)}{x-c} \right)} = \frac{f'(c)}{g'(c)} = \lim_{x \to c} \frac{f'(x)}{g'(x)}.$$

This follows from the difference-quotient definition of the derivative. The last equality follows from the continuity of the derivatives at c. The limit in the conclusion is not indeterminate because $g'(c) \neq 0$.

Proof:

The following proof is due to Taylor, where a unified proof for the 0/0 and $\pm\infty/\pm\infty$ indeterminate forms is given. Taylor notes that different proofs may be found in Lettenmeyer and Wazewski.

Let f and g be functions satisfying the hypotheses in the General form section. Let $\mathcal{I}$ be the open interval in the hypothesis with endpoint c. Considering that $g'(x) \neq 0$ on this interval and g is continuous, $\mathcal{I}$ can be chosen smaller so that g is nonzero on $\mathcal{I}$.

For each x in the interval, define $m(x) = \inf \dfrac{f'(\xi)}{g'(\xi)}$ and $M(x) = \sup \dfrac{f'(\xi)}{g'(\xi)}$ as ξ ranges over all values between x and c. (The symbols inf and sup denote the infimum and supremum.)

From the differentiability of f and g on $\mathcal{I}$, Cauchy's mean value theorem ensures that for any two distinct points x and y in $\mathcal{I}$ there exists a ξ between x and y such that $\dfrac{f(x) - f(y)}{g(x) - g(y)} = \dfrac{f'(\xi)}{g'(\xi)}$.

Consequently, $m(x) \leq \dfrac{f(x) - f(y)}{g(x) - g(y)} \leq M(x)$ for all choices of distinct x and y in the interval. The value $g(x)$-$g(y)$ is always nonzero for distinct x and y in the interval, for if it was not, the mean value theorem would imply the existence of a p between x and y such that $g'(p)$=0.

The definition of $m(x)$ and $M(x)$ will result in an extended real number, and so it is possible for them to take on the values $\pm\infty$. In the following two cases, $m(x)$ and $M(x)$ will establish bounds on the ratio f

Case 1: $\lim\limits_{x \to c} f(x) = \lim\limits_{x \to c} g(x) = 0$

For any x in the interval $\mathcal{I}$, and point y between x and c,

$$m(x) \leq \frac{f(x) - f(y)}{g(x) - g(y)} = \frac{\dfrac{f(x)}{g(x)} - \dfrac{f(y)}{g(x)}}{1 - \dfrac{g(y)}{g(x)}} \leq M(x)$$

and therefore as y approaches c, $\dfrac{f(y)}{g(x)}$ and $\dfrac{g(y)}{g(x)}$ become zero, and so

$$m(x) \leq \frac{f(x)}{g(x)} \leq M(x).$$

Case 2: $\lim\limits_{x \to c} | g(x) | = \infty$

For every x in the interval $\mathcal{I}$, define $S_x = \{y \mid y \text{ is between } x \text{ and } c\}$. For every point y between x and c,

$$m(x) \leq \frac{f(y) - f(x)}{g(y) - g(x)} = \frac{\dfrac{f(y)}{g(y)} - \dfrac{f(x)}{g(y)}}{1 - \dfrac{g(x)}{g(y)}} \leq M(x).$$

As y approaches c, both $\dfrac{f(x)}{g(y)}$ and $\dfrac{g(x)}{g(y)}$ become zero and therefore,

$$m(x) \leq \liminf_{y \in S_x} \frac{f(y)}{g(y)} \leq \limsup_{y \in S_x} \frac{f(y)}{g(y)} \leq M(x).$$

The limit superior and limit inferior are necessary since the existence of the limit of f/g has not yet been established.

It is also the case that,

$$\lim_{x \to c} m(x) = \lim_{x \to c} M(x) = \lim_{x \to c} \frac{f'(x)}{g'(x)} = L.$$

and

$$\lim_{x \to c}\left(\liminf_{y \in S_x} \frac{f(y)}{g(y)} \right) = \liminf_{x \to c} \frac{f(x)}{g(x)} \quad \text{and} \quad \lim_{x \to c}\left(\limsup_{y \in S_x} \frac{f(y)}{g(y)} \right) = \limsup_{x \to c} \frac{f(x)}{g(x)}.$$

In case 1, the squeeze theorem establishes that $\lim\limits_{x \to c} \dfrac{f(x)}{g(x)}$ exists and is equal to L. In the case 2, and

the squeeze theorem again asserts that $\liminf\limits_{x \to c} \dfrac{f(x)}{g(x)} = \limsup\limits_{x \to c} \dfrac{f(x)}{g(x)} = L$, and so the limit $\lim\limits_{x \to c} \dfrac{f(x)}{g(x)}$

exists and is equal to L. This is the result that was to be proven.

In case 2 the assumption that $f(x)$ diverges to infinity was not used within the proof. This means that if $|g(x)|$ diverges to infinity as x approaches c and both f and g satisfy the hypotheses of L'Hôpital's rule, then no additional assumption is needed about the limit of $f(x)$: It could even be the case that the limit of $f(x)$ does not exist. In this case, L'Hopital's theorem is actually a consequence of Cesàro–Stolz.

In the case when $|g(x)|$ diverges to infinity as x approaches c and $f(x)$ converges to a finite limit at c, then L'Hôpital's rule would be applicable, but not absolutely necessary, since basic limit calculus will show that the limit of $f(x)/g(x)$ as x approaches c must be zero.

Corollary

A simple but very useful consequence of L'Hopital's rule is a well-known criterion for differentiability. It states the following: suppose that f is continuous at a, and that $f'(x)$ exists for all x in some open interval containing a, except perhaps for $x = a$. Suppose, moreover, that $\lim\limits_{x \to a} f'(x)$ exists. Then $f'(a)$ also exists and,

$$f'(a) = \lim_{x \to a} f'(x).$$

In particular, f is also continuous at a.

Proof:

Consider the functions $h(x) = f(x) - f(a)$ and $g(x) = x - a$. The continuity of f at a tells us that $\lim_{x \to a} h(x) = 0$. Moreover, $\lim_{x \to a} g(x) = 0$ since a polynomial function is always continuous everywhere. Applying L'Hopital's rule shows that $f'(a) := \lim_{x \to a} \dfrac{f(x) - f(a)}{x - a} = \lim_{x \to a} \dfrac{h(x)}{g(x)} = \lim_{x \to a} f'(x)$.

EXTREME VALUE THEOREM

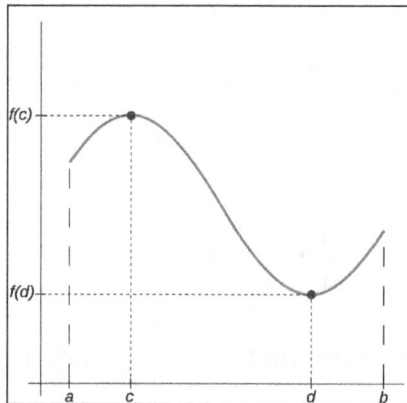

A continuous function $f(x)$ on the closed interval [a,b] showing the absolute max (red) and the absolute min (blue).

In calculus, the extreme value theorem states that if a real-valued function f is continuous on the closed interval [a,b], then f must attain a maximum and a minimum, each at least once. That is, there exist numbers c and d in [a,b] such that:

$$f(c) \geq f(x) \geq f(d) \quad \text{for all } x \in [a,b]$$

A related theorem is the boundedness theorem which states that a continuous function f in the closed interval [a,b] is bounded on that interval. That is, there exist real numbers m and M such that:

$$m < f(x) < M \quad \text{for all } x \in [a,b]$$

The extreme value theorem enriches the boundedness theorem by saying that not only is the function bounded, but it also attains its least upper bound as its maximum and its greatest lower bound as its minimum.

The extreme value theorem is used to prove Rolle's theorem. In a formulation due to Karl Weierstrass, this theorem states that a continuous function from a non-empty compact space to a subset of the real numbers attains a maximum and a minimum.

Functions to which the Theorem does not Apply

The following examples show why the function domain must be closed and bounded in order for the theorem to apply. Each fails to attain a maximum on the given interval.

1. $f(x) = x$ defined over $[0, \infty)$ is not bounded from above.

2. $f(x) = x/(1 + x)$ defined over $[0, \infty)$ is bounded but does not attain its least upper bound 1.

3. $f(x) = 1/x$ defined over $(0, 1]$ is not bounded from above.

4. $f(x) = 1 - x$ defined over $(0, 1]$ is bounded but never attains its least upper bound 1.

Defining $f(0) = 0$ in the last two examples shows that both theorems require continuity on $[a, b]$.

Generalization to Metric and Topological Spaces

When moving from the real line $\mathbb{R}$ to metric spaces and general topological spaces, the appropriate generalization of a closed bounded interval is a compact set. A set K is said to be compact if it has the following property: from every collection of open sets U_α such that $\bigcup U_\alpha \supset K$, a finite subcollection $U_{\alpha_1}, \ldots, U_{\alpha_n}$ can be chosen such that $\bigcup_{i=1}^{n} U_{\alpha_i} \supset K$. This is called the Heine–Borel property, and it is usually stated in short as "every open cover of K has a finite subcover". The Heine–Borel theorem asserts that a subset of the real line is compact if and only if it is both closed and bounded.

The concept of a continuous function can likewise be generalized. Given topological spaces V, W, a function $f : V \to W$ is said to be continuous if for every open set $U \subset W$, $f^{-1}(U) \subset V$ is also open. Given these definitions, continuous functions can be shown to preserve compactness:

Theorem: *If V, W are topological spaces, $f : V \to W$ is a continuous function, and $K \subset V$ is compact, then $f(K) \subset W$ is also compact.*

In particular, if $W = \mathbb{R}$, then this theorem implies that $f(K)$ is closed and bounded for any compact set K, which in turn implies that f attains its supremum and infimum on any (nonempty) compact set K. Thus, we have the following generalization of the extreme value theorem:

Theorem: *If K is a compact set and $f : K \to \mathbb{R}$ is a continuous function, then f is bounded and there exist $p, q \in K$ such that $f(p) = \sup_{x \in K} f(x)$ and $f(q) = \inf_{x \in K} f(x)$.*

Slightly more generally, this is also true for an upper semicontinuous function.

Proving the Theorems

We look at the proof for the upper bound and the maximum of f. By applying these results to the function $-f$, the existence of the lower bound and the result for the minimum of f follows. Also note that everything in the proof is done within the context of the real numbers.

We first prove the boundedness theorem, which is a step in the proof of the extreme value theorem. The basic steps involved in the proof of the extreme value theorem are:

1. Prove the boundedness theorem.

2. Find a sequence so that its image converges to the supremum of f.

3. Show that there exists a subsequence that converges to a point in the domain.

4. Use continuity to show that the image of the subsequence converges to the supremum.

Proof of the Boundedness Theorem

Statement: If $f(x)$ is continuous on $[a,b]$ then it is bounded on $[a,b]$.

Suppose the function f is not bounded above on the interval $[a,b]$. Then, for every natural number n, there exists an x_n in $[a,b]$ such that $f(x_n) > n$. This defines a sequence $\{x_n\}$. Because $[a,b]$ is bounded, the Bolzano–Weierstrass theorem implies that there exists a convergent subsequence $\{x_{n_k}\}$ of $\{x_n\}$. Denote its limit by x. As $[a,b]$ is closed, it contains x. Because f is continuous at x, we know that $\{f(x_{n_k})\}$ converges to the real number $f(x)$ (as f is sequentially continuous at x.) But $f(x_{nk}) > n_k \geq k$ for every k, which implies that $\{f(x_{nk})\}$ diverges to $+\infty$, a contradiction. Therefore, f is bounded above on $[a,b]$.

Alternative Proof

Statement: If $f(x)$ is continuous on $[a,b]$ then it is bounded on $[a,b]$.

Proof Consider the set B of points x in $[a,b]$ such that $f(x)$ is bounded on $[a,x]$. We note that a is one such point, for $f(x)$ is bounded on $[a,a]$ by the value $f(a)$. If $e > a$ is another point, then all points between a and e also belong to B. In other words B is an interval closed at its left end by a.

Now f is continuous on the right at a, hence there exists $\delta > 0$ such that $|f(x) - f(a)| < 1$ for all x in $[a, a+\delta]$. Thus f is bounded by $f(a) - 1$ and $f(a) + 1$ on the interval $[a, a+\delta]$ so that all these points belong to B.

So far, we know that B is an interval of non-zero length, closed at its left end by a.

Next, B is bounded above by b. Hence the set B has a supremum in $[a,b]$; let us call it s. From the non-zero length of B we can deduce that $s > a$.

Suppose $s < b$. Now f is continuous at s, hence there exists $\delta > 0$ such that $|f(x) - f(s)| < 1$ for all x in $[s - \delta, s + \delta]$ so that f is bounded on this interval. But it follows from the supremacy of s that there exists a point belonging to B, e say, which is greater than $s - \delta / 2$. Thus f is bounded on $[a,e]$ which overlaps $[s-\delta, s+\delta]$ so that f is bounded on $[a, s+\delta]$. This however contradicts the supremacy of s.

We must therefore have $s = b$. Now f is continuous on the left at s, hence there exists $\delta > 0$ such that $|f(x) - f(s)| < 1$ for all x in $[s-\delta, s]$ so that f is bounded on this interval. But it follows from the supremacy of s that that there exists a point belonging to B, e say, which is greater than $s - \delta / 2$. Thus f is bounded on $[a,e]$ which overlaps $[s-\delta, s]$ so that f is bounded on $[a,s]$.

Proof of the Extreme Value Theorem

By the boundedness theorem, f is bounded from above, hence, by the Dedekind-completeness of the real numbers, the least upper bound (supremum) M of f exists. It is necessary to find a point d

in $[a,b]$ such that $M = f(d)$. Let n be a natural number. As M is the *least* upper bound, $M - 1/n$ is not an upper bound for f. Therefore, there exists d_n in $[a,b]$ so that $M - 1/n < f(d_n)$. This defines a sequence $\{d_n\}$. Since M is an upper bound for f, we have $M - 1/n < f(d_n) \leq M + 1/n$ for all n. Therefore, the sequence $\{f(d_n)\}$ converges to M.

The Bolzano–Weierstrass theorem tells us that there exists a subsequence $\{d_{n_k}\}$, which converges to some d and, as $[a,b]$ is closed, d is in $[a,b]$. Since f is continuous at d, the sequence $\{f(d_{n_k})\}$ converges to $f(d)$. But $\{f(d_{nk})\}$ is a subsequence of $\{f(d_n)\}$ that converges to M, so $M = f(d)$. Therefore, f attains its supremum M at d.

Alternative Proof of the Extreme Value Theorem

The set $\{y \in R : y = f(x) \text{ for some } x \in [a,b]\}$ is a bounded set. Hence, its least upper bound exists by least upper bound property of the real numbers. Let $M = \sup(f(x))$ on $[a, b]$. If there is no point x on $[a, b]$ so that $f(x) = M$ then $f(x) < M$ on $[a, b]$. Therefore, $1/(M - f(x))$ is continuous on $[a, b]$.

However, to every positive number ε, there is always some x in $[a, b]$ such that $M - f(x) < \varepsilon$ because M is the least upper bound. Hence, $1/(M - f(x)) > 1/\varepsilon$, which means that $1/(M - f(x))$ is not bounded. Since every continuous function on a $[a, b]$ is bounded, this contradicts the conclusion that $1/(M - f(x))$ was continuous on $[a, b]$. Therefore, there must be a point x in $[a, b]$ such that $f(x) = M$.

Proof using the Hyperreals

In the setting of non-standard calculus, let N be an infinite hyperinteger. The interval $[0, 1]$ has a natural hyperreal extension. Consider its partition into N subintervals of equal infinitesimal length $1/N$, with partition points $x_i = i/N$ as i "runs" from 0 to N. The function f is also naturally extended to a function f^* defined on the hyperreals between 0 and 1. Note that in the standard setting (when N is finite), a point with the maximal value of f can always be chosen among the $N+1$ points x_i, by induction. Hence, by the transfer principle, there is a hyperinteger i_0 such that $0 \leq i_0 \leq N$ and $f^*(x_{i_0}) \geq f^*(x_i)$ for all $i = 0, ..., N$. Consider the real point,

$$c = \text{st}(x_{i_0})$$

where st is the standard part function. An arbitrary real point x lies in a suitable sub-interval of the partition, namely $x \in [x_i, x_{i+1}]$, so that $\text{st}(x_i) = x$. Applying st to the inequality $f^*(x_{i_0}) \geq f^*(x_i)$, we obtain $\text{st}(f^*(x_{i_0})) \geq \text{st}(f^*(x_i))$. By continuity of f we have,

$$\text{st}(f^*(x_{i_0})) = f(\text{st}(x_{i_0})) = f(c).$$

Hence $f(c) \geq f(x)$, for all real x, proving c to be a maximum of f.

Proof from First Principles

Statement: If $f(x)$ is continuous on $[a,b]$ then it attains its supremum on $[a,b]$.

Proof By the Boundedness Theorem, $f(x)$ is bounded above on $[a,b]$ and by the completeness property of the real numbers has a supremum in $[a,b]$. Let us call it M, or $M[a,b]$. It is clear that the restriction of f to the subinterval $[a,x]$ where $x \leq b$ has a supremum $M[a,x]$ which is less than or equal to M, and that $M[a,x]$ increases from $f(a)$ to M as x increases from a to b.

If $f(a) = M$ then we are done. Suppose therefore that $f(a) < M$ and let $d = M - f(a)$. Consider the set L of points x in $[a,b]$ such that $M[a,x] < M$.

Clearly $a \in L$; moreover if $e > a$ is another point in L then all points between a and e also belong to L because $M[a,x]$ is monotonic increasing. Hence L is a non-empty interval, closed at its left end by a.

Now f is continuous on the right at a, hence there exists $\delta > 0$ such that $|f(x) - f(a)| < d/2$ for all x in $[a, a+\delta]$. Thus f is less than $M - d/2$ on the interval $[a, a+\delta]$ so that all these points belong to L.

Next, L is bounded above by b and has therefore a supremum in $[a,b]$: let us call it s. We see from the above that $s \quad a$. We will show that s is the point we are seeking i.e. the point where f attains its supremum, or in other words $f(s) = M$.

Suppose the contrary viz. $f(s) < M$. Let $d = M - f(s)$ and consider the following two cases:

(1) $s < b$. As f is continuous at s, there exists $\delta > 0$ such that $|f(x) - f(s)| < d/2$ for all x in $[s-\delta, s+\delta]$. This means that f is less than $M - d/2$ on the interval $[s-\delta, s+\delta]$. But it follows from the supremacy of s that there exists a point, e say, belonging to L which is greater than $s - \delta$. By the definition of L, $M[a,e] < M$. Let $d_1 = M - M[a,e]$ then for all x in $[a,e]$, $f(x) \leq M - d_1$. Taking d_2 to be the minimum of $d/2$ and d_1, we have $f(x) \leq M - d_2$ for all x in $[a, s+\delta]$.

Hence $M[a, s+\delta] < M$ so that $s + \delta \in L$. This however contradicts the supremacy of s and completes the proof.

(2) $s = b$. As f is continuous on the left at s, there exists $\delta > 0$ such that $|f(x) - f(s)| < d/2$ for all x in $[s-\delta, s]$. This means that f is less than $M - d/2$ on the interval $[s-\delta, s]$. But it follows from the supremacy of s that there exists a point, e say, belonging to L which is greater than $s - \delta$. By the definition of L, $M[a,e] < M$. Let $d_1 = M - M[a,e]$ then for all x in $[a,e]$, $f(x) \leq M - d_1$. Taking d_2 to be the minimum of $d/2$ and d_1, we have $f(x) \leq M - d_2$ for all x in $[a,b]$. This contradicts the supremacy of M and completes the proof.

Extension to Semi-continuous Functions

If the continuity of the function f is weakened to semi-continuity, then the corresponding half of the boundedness theorem and the extreme value theorem hold and the values $-\infty$ or $+\infty$, respectively, from the extended real number line can be allowed as possible values. More precisely.

Theorem: If a function $f: [a,b] \to [-\infty, \infty)$ is upper semi-continuous, meaning that,

$$\limsup_{y \to x} f(y) \leq f(x)$$

for all x in $[a,b]$, then f is bounded above and attains its supremum.

Proof: If $f(x) = -\infty$ for all x in $[a,b]$, then the supremum is also $-\infty$ and the theorem is true. In all other cases, the proof is a slight modification of the proofs given above. In the proof of the boundedness theorem, the upper semi-continuity of f at x only implies that the limit superior of the subsequence $\{f(x_{n_k})\}$ is bounded above by $f(x) < \infty$, but that is enough to obtain the contradiction. In the proof of the extreme value theorem, upper semi-continuity of f at d implies that the limit superior of the subsequence $\{f(d_{n_k})\}$ is bounded above by $f(d)$, but this suffices to conclude that $f(d) = M$.

Applying this result to $-f$ proves:

Theorem: If a function $f: [a,b] \to (-\infty, \infty]$ is lower semi-continuous, meaning that,

$$\liminf_{y \to x} f(y) \geq f(x)$$

for all x in $[a,b]$, then f is bounded below and attains its infimum.

A real-valued function is upper as well as lower semi-continuous, if and only if it is continuous in the usual sense. Hence these two theorems imply the boundedness theorem and the extreme value theorem.

IMPLICIT FUNCTION THEOREM

In mathematics, more specifically in multivariable calculus, the implicit function theorem is a tool that allows relations to be converted to functions of several real variables. It does so by representing the relation as the graph of a function. There may not be a single function whose graph can represent the entire relation, but there may be such a function on a restriction of the domain of the relation. The implicit function theorem gives a sufficient condition to ensure that there is such a function.

More precisely, given a system of m equations $f_i(x_1, \ldots, x_n, y_1, \ldots, y_m) = 0$, $i = 1, \ldots, m$ (often abbreviated into $F(x, y) = 0$), the theorem states that, under a mild condition on the partial derivatives (with respect to the y_is) at a point, the m variables y_i are differentiable functions of the x_j in some neighborhood of the point. As these functions can generally not be expressed in closed form, they are *implicitly* defined by the equations, and this motivated the name of the theorem.

In other words, under a mild condition on the partial derivatives, the set of zeros of a system of equations is locally the graph of a function.

Example:

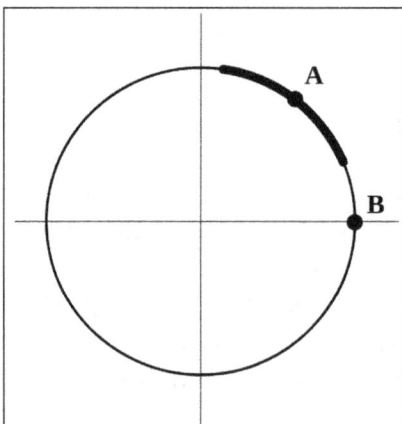

The unit circle can be specified as the level curve $f(x, y) = 1$ of the function $f(x, y) = x^2 + y^2$ Around point A, y can be expressed as a function $y(x)$. In this example this function can be written explicitly as $g_1(x) = \sqrt{1 - x^2}$ in many cases no such explicit expression exists, but one can still refer to the *implicit* function $y(x)$. No such function exists around point B.

If we define the function $f(x, y) = x^2 + y^2$, then the equation $f(x, y) = 1$ cuts out the unit circle as the level set $\{(x, y)\,|\, f(x, y) = 1\}$. There is no way to represent the unit circle as the graph of a function of one variable $y = g(x)$ because for each choice of $x \in (-1, 1)$, there are two choices of y, namely $\pm\sqrt{1 - x^2}$.

However, it is possible to represent *part* of the circle as the graph of a function of one variable. If we let $g_1(x) = \sqrt{1 - x^2}$ for $-1 \le x \le 1$, then the graph of $y = g_1(x)$ provides the upper half of the circle. Similarly, if $g_2(x) = -\sqrt{1 - x^2}$, then the graph of $y = g_2(x)$ gives the lower half of the circle.

The purpose of the implicit function theorem is to tell us the existence of functions like $g_1(x)$ and $g_2(x)$, even in situations where we cannot write down explicit formulas. It guarantees that $g_1(x)$ and $g_2(x)$ are differentiable, and it even works in situations where we do not have a formula for $f(x, y)$.

Let $f: R^{n+m} \to R^m$ be a continuously differentiable function. We think of R^{n+m} as the Cartesian product $R^n \times R^m$, and we write a point of this product as $(x, y) = (x_1, ..., x_n, y_1, ..., y_m)$. Starting from the given function f, our goal is to construct a function $g: R^n \to R^m$ whose graph $(x, g(x))$ is precisely the set of all (x, y) such that $f(x, y) = 0$.

As noted above, this may not always be possible. We will therefore fix a point $(a, b) = (a_1, ..., a_n, b_1, ..., b_m)$ which satisfies $f(a, b) = 0$, and we will ask for a g that works near the point (a, b). In other words, we want an open set U of R^n containing a, an open set V of R^m containing b, and a function $g : U \to V$ such that the graph of g satisfies the relation $f = 0$ on $U \times V$, and that no other points within $U \times V$ do so. In symbols,

$$\{(x, g(x))\,|\, x \in U\} = \{(x, y) \in U \times V \,|\, f(x, y) = 0\}.$$

To state the implicit function theorem, we need the Jacobian matrix of f, which is the matrix of the partial derivatives of f. Abbreviating $(a_1, ..., a_n, b_1, ..., b_m)$ to (a, b), the Jacobian matrix is,

$$(Df)(a,b) = \begin{bmatrix} \dfrac{\partial f_1}{\partial x_1}(a,b) & \cdots & \dfrac{\partial f_1}{\partial x_n}(a,b) & \dfrac{\partial f_1}{\partial y_1}(a,b) & \cdots & \dfrac{\partial f_1}{\partial y_m}(a,b) \\ \vdots & \ddots & \vdots & \vdots & \ddots & \vdots \\ \dfrac{\partial f_m}{\partial x_1}(a,b) & \cdots & \dfrac{\partial f_m}{\partial x_n}(a,b) & \dfrac{\partial f_m}{\partial y_1}(a,b) & \cdots & \dfrac{\partial f_m}{\partial y_m}(a,b) \end{bmatrix} = [X \,|\, Y]$$

where X is the matrix of partial derivatives in the variables x_i and Y is the matrix of partial derivatives in the variables y_j. The implicit function theorem says that if Y is an invertible matrix, then there are U, V, and g as desired. Writing all the hypotheses together gives the following statement.

Statement of the Theorem

Let f: $\mathbb{R}^{n+m} \to \mathbb{R}^m$ be a continuously differentiable function, and let $\mathbb{R}^{n+m}$ have coordinates (x, y). Fix a point $(a, b) = (a_1, ..., a_n, b_1, ..., b_m)$ with $f(a, b) = 0$, where $0 \in \mathbb{R}^m$ is the zero vector. If the Jacobian matrix $J_{f,y}(a, b) = [(\partial f_i / \partial y_j)(a, b)]$ is invertible, then there exists an open set U of $\mathbb{R}^n$ containing a such that there exists a unique continuously differentiable function g: $U \to \mathbb{R}^m$ such that,

$$g(a) = b$$

and

$$f(x, g(x)) = 0 \quad \text{for all} \quad x \in U.$$

Moreover, the partial derivatives of g in U are given by the matrix product,

$$\frac{\partial g}{\partial x_j}(x) = -\Big[J_{f,y}(x, g(x)) \Big]_{m \times m}^{-1} \left[\frac{\partial f}{\partial x_j}(x, g(x)) \right]_{m \times 1}.$$

Higher Derivatives

If, moreover, f is analytic or continuously differentiable k times in a neighborhood of (a, b), then one may choose U in order that the same holds true for g inside U. In the analytic case, this is called the analytic implicit function theorem.

Proof for 2D Case

Let us assume that F: $\mathbb{R}^2 \to \mathbb{R}$ is a continuously differentiable function defining a curve $F(x, y) = 0$. Let (x_o, y_o) be a point on the curve, that is, $F(x_o, y_o) = 0$. The statement of the theorem above can be rewritten for this simple case as follows:

If $\dfrac{\partial F}{\partial y}\Big|_{(x_0, y_0)} \neq 0$ then for the curve around (x_o, y_o) we can write $y = f(x)$, where f is a real function.

Proof: Since F is differentiable we write the differential of F through partial derivatives:

$dF = \text{grad} F \cdot d\vec{x} = \dfrac{\partial F}{\partial x} dx + \dfrac{\partial F}{\partial y} dy$. Since we are restricted to movement on the curve $dF = 0$ and by assumption $\dfrac{\partial F}{\partial y} \neq 0$ around the point (x_o, y_o). Therefore we have a first-order ordinary differential equation:

$$\partial_x F \, dx + \partial_y F \, dy = 0, \quad y(x_0) = y_0$$

Now we are looking for a solution to this ODE in an open interval around the point (x_o, y_o) for which, at every point in it, $\partial_y F \neq 0$. Since F is continuously differentiable and from the assumption we have $|\partial_x F| < \infty, |\partial_y F| < \infty, \partial_y F \neq 0$. From this we know that $\dfrac{\partial_x F}{\partial_y F}$ is continuous and bounded on both ends. From here we know that $-\dfrac{\partial_x F}{\partial_y F}$ is Lipschitz continuous in both x and y. Therefore, by Cauchy-Lipschitz theorem, there exists unique $y(x)$ that is the solution to the given ODE with the initial conditions.

The Circle Example

Let us go back to the example of the unit circle. In this case $n = m = 1$ and $f(x, y) = x^2 + y^2 - 1$. The matrix of partial derivatives is just a 1×2 matrix given by:

$$(Df)(a,b) = \left[\frac{\partial f}{\partial x}(a,b) \ \ \frac{\partial f}{\partial y}(a,b) \right] = [2a \ \ 2b]$$

Thus, here, the Y in the statement of the theorem is just the number $2b$; the linear map defined by it is invertible iff $b \neq 0$. By the implicit function theorem we see that we can locally write the circle in the form $y = g(x)$ for all points where $y \neq 0$. For $(\pm 1, 0)$ we run into trouble, as noted before. The implicit function theorem may still be applied to these two points, by writing x as a function of y, that is, $x = h(y)$; now the graph of the function will be $\big(h(y), y\big)$, since where $b = 0$ we have $a = 1$, and the conditions to locally express the function in this form are satisfied.

The implicit derivative of y with respect to x, and that of x with respect to y, can be found by totally differentiating the implicit function $x^2 + y^2 - 1$ and equating to 0:

$$2x dx + 2y dy = 0$$

giving,

$$dy / dx = -x / y,$$

and

$$dx / dy = -y / x.$$

Application: Change of Coordinates

Suppose we have an m-dimensional space, parametrised by a set of coordinates $(x_1, \ldots, x_m)$. We can introduce a new coordinate system $(x'_1, \ldots, x'_m)$ by supplying m functions $h_1 \ldots h_m$ each being continuously differentiable. These functions allow us to calculate the new coordinates $(x'_1, \ldots, x'_m)$ of a point, given the point's old coordinates $(x_1,\ldots,x_m)$ using $x'_1 = h_1(x_1,\ldots,x_m),\ldots,x'_m = h_m(x_1,\ldots,x_m)$. One might want to verify if the opposite is possible: given coordinates $(x'_1,\ldots,x'_m)$, can we 'go back' and calculate the same point's original coordinates $(x_1,\ldots,x_m)$? The implicit function theorem will provide an answer to this question. The (new and old) coordinates $(x'_1,\ldots,x'_m,x_1,\ldots,x_m)$ are related by $f = 0$, with

$$f(x'_1,\ldots,x'_m,x_1,\ldots,x_m) = (h_1(x_1,\ldots,x_m) - x'_1,\ldots,h_m(x_1,\ldots,x_m) - x'_m).$$

Now the Jacobian matrix of f at a certain point (a, b) [where $a = (x'_1,\ldots,x'_m), b = (x_1,\ldots,x_m)$] is given by:

$$(Df)(a,b) = \left[\begin{array}{ccc|ccc} -1 & \cdots & 0 & \dfrac{\partial h_1}{\partial x_1}(b) & \cdots & \dfrac{\partial h_1}{\partial x_m}(b) \\ \vdots & \ddots & \vdots & \vdots & \ddots & \vdots \\ 0 & \cdots & -1 & \dfrac{\partial h_m}{\partial x_1}(b) & \cdots & \dfrac{\partial h_m}{\partial x_m}(b) \end{array}\right] = [-I_m \,|\, J].$$

where I_m denotes the $m \times m$ identity matrix, and J is the $m \times m$ matrix of partial derivatives, evaluated at (a, b). (In the above, these blocks were denoted by X and Y. As it happens, in this particular application of the theorem, neither matrix depends on a.) The implicit function theorem now states that we can locally express $(x_1,\ldots,x_m)$ as a function of $(x'_1,\ldots,x'_m)$ if J is invertible. Demanding J is invertible is equivalent to $\det J \neq 0$, thus we see that we can go back from the primed to the unprimed coordinates if the determinant of the Jacobian J is non-zero. This statement is also known as the inverse function theorem.

Example: Polar Coordinates

As a simple application of the above, consider the plane, parametrised by polar coordinates (R, θ). We can go to a new coordinate system (cartesian coordinates) by defining functions $x(R, \theta) = R \cos(\theta)$ and $y(R, \theta) = R \sin(\theta)$. This makes it possible given any point (R, θ) to find corresponding cartesian coordinates (x, y). When can we go back and convert cartesian into polar coordinates? By the previous example, it is sufficient to have $\det J \neq 0$, with

$$J = \begin{bmatrix} \dfrac{\partial x(R,\theta)}{\partial R} & \dfrac{\partial x(R,\theta)}{\partial \theta} \\ \dfrac{\partial y(R,\theta)}{\partial R} & \dfrac{\partial y(R,\theta)}{\partial \theta} \end{bmatrix} = \begin{bmatrix} \cos\theta & -R\sin\theta \\ \sin\theta & R\cos\theta \end{bmatrix}.$$

Since $\det J = R$, conversion back to polar coordinates is possible if $R \neq 0$. So it remains to check the case $R = 0$. It is easy to see that in case $R = 0$, our coordinate transformation is not invertible: at the origin, the value of θ is not well-defined.

Generalizations

Banach Space Version

Based on the inverse function theorem in Banach spaces, it is possible to extend the implicit function theorem to Banach space valued mappings.

Let X, Y, Z be Banach spaces. Let the mapping $f : X \times Y \to Z$ be continuously Fréchet differentiable. If $(x_0, y_0) \in X \times Y$, $f(x_0, y_0) = 0$, and $y \mapsto Df(x_0, y_0)(0, y)$ is a Banach space isomorphism from Y onto Z, then there exist neighbourhoods U of x_0 and V of y_0 and a Fréchet differentiable function $g :$ $U \to V$ such that $f(x, g(x)) = 0$ and $f(x, y) = 0$ if and only if $y = g(x)$, for all $(x, y) \in U \times V$.

Implicit Functions from Non-differentiable Functions

Various forms of the implicit function theorem exist for the case when the function f is not differentiable. It is standard that local strict monotonicity suffices in one dimension. The following more general form was proven by Kumagai based on an observation by Jittorntrum.

Consider a continuous function $f : R^n \times R^m \to R^n$ such that $f(x_0, y_0) = 0$. There exist open neighbourhoods $A \subset R^n$ and $B \subset R^m$ of x_0 and y_0, respectively, such that, for all y in B, $f(\cdot, y) : A \to R^n$ is locally one-to-one *if and only if* there exist open neighbourhoods $A_0 \subset R^n$ and $B_0 \subset R^m$ of x_0 and y_0, such that, for all $y \in B_0$, the equation $f(x, y) = 0$ has a unique solution:

$$x = g(y) \in A_0,$$

where g is a continuous function from B_0 into A_0.

TAYLOR'S THEOREM

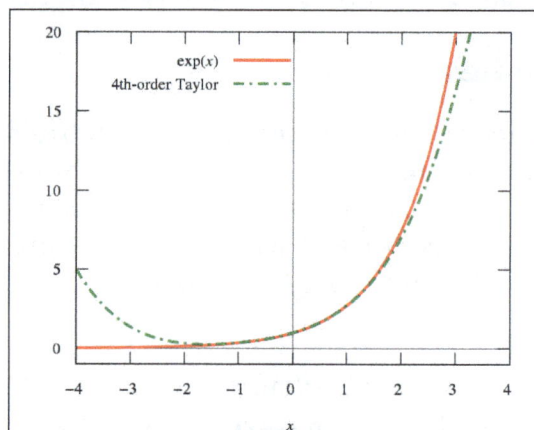

The exponential function $y = e^x$ (red) and the corresponding Taylor polynomial of degree four (dashed green) around the origin.

In calculus, Taylor's theorem gives an approximation of a k-times differentiable function around a given point by a k-th order Taylor polynomial. For analytic functions the Taylor polynomials at a given point are finite-order truncations of its Taylor series, which completely determines the function in some neighborhood of the point. It can be thought of as the extension of linear approximation to higher order polynomials, and in the case of k equals 2 is often referred to as a *quadratic approximation*. The exact content of "Taylor's theorem" is not universally agreed upon. Indeed, there are several versions of it applicable in different situations, and some of them contain explicit estimates on the approximation error of the function by its Taylor polynomial.

Taylor's theorem is named after the mathematician Brook Taylor, who stated a version of it in 1712. Yet an explicit expression of the error was not provided until much later on by Joseph-Louis Lagrange. An earlier version of the result was already mentioned in 1671 by James Gregory.

Taylor's theorem is taught in introductory-level calculus courses and is one of the central elementary tools in mathematical analysis. Within pure mathematics it is the starting point of more advanced asymptotic analysis and is commonly used in more applied fields of numerics, as well as in mathematical physics. Taylor's theorem also generalizes to multivariate and vector valued functions $f : \mathbb{R}^n \to \mathbb{R}^m$ on any dimensions n and m. This generalization of Taylor's theorem is the basis for the definition of so-called jets, which appear in differential geometry and partial differential equations.

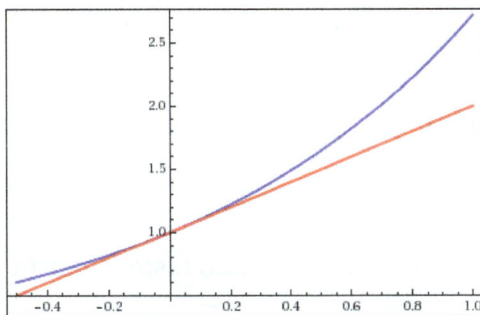

Graph of $f(x) = e^x$ (blue) with its linear approximation $P_1(x) = 1 + x$ (red) at $a = 0$.

If a real-valued function f is differentiable at the point a then it has a linear approximation at the point a. This means that there exists a function h_1 such that,

$$f(x) = f(a) + f'(a)(x - a) + h_1(x)(x - a), \quad \lim_{x \to a} h_1(x) = 0.$$

Here,

$$P_1(x) = f(a) + f'(a)(x - a)$$

is the linear approximation of f at the point a. The graph of $y = P_1(x)$ is the tangent line to the graph of f at $x = a$. The error in the approximation is:

$$R_1(x) = f(x) - P_1(x) = h_1(x)(x - a).$$

This goes to zero a little bit faster than $x - a$ as x tends to a, given the limiting behavior of h_1.

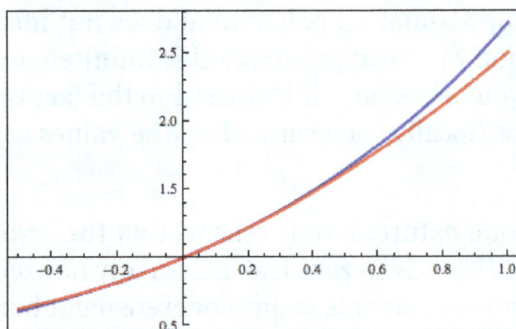

Graph of $f(x) = e^x$ (blue) with its quadratic approximation $P_2(x) = 1 + x + x^2/2$ (red) at $a = 0$.

If we wanted a better approximation to f, we might instead try a quadratic polynomial instead of a linear function. Instead of just matching one derivative of f at a, we can match two derivatives, thus producing a polynomial that has the same slope and concavity as f at a. The quadratic polynomial in question is:

$$P_2(x) = f(a) + f'(a)(x-a) + \frac{f''(a)}{2}(x-a)^2.$$

Taylor's theorem: Ensures that the *quadratic approximation* is, in a sufficiently small neighborhood of the point a, a better approximation than the linear approximation. Specifically,

$$f(x) = P_2(x) + h_2(x)(x-a)^2, \quad \lim_{x \to a} h_2(x) = 0.$$

Here the error in the approximation is,

$$R_2(x) = f(x) - P_2(x) = h_2(x)(x-a)^2,$$

which, given the limiting behavior of h_2, goes to zero faster than $(x-a)^2$ as x tends to a.

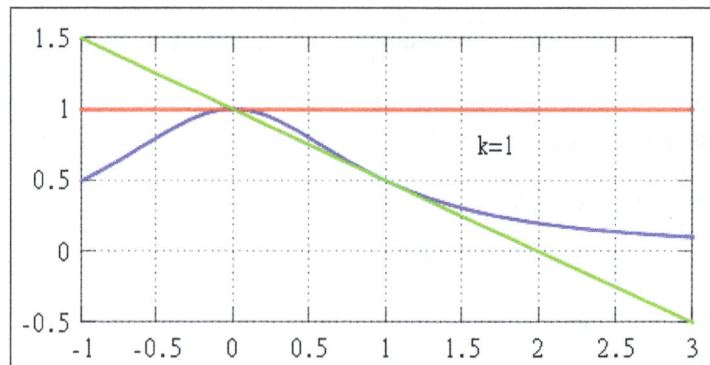

Approximation of $f(x) = 1/(1+x^2)$ (blue) by its Taylor polynomials P_k of order $k = 1, ..., 16$ centered at $x = 0$ (red) and $x = 1$ (green). The approximations do not improve at all outside $(-1, 1)$ and $(1 - \sqrt{2}, 1 + \sqrt{2})$ respectively.

Similarly, we might get still better approximations to f if we use polynomials of higher degree, since then we can match even more derivatives with f at the selected base point.

In general, the error in approximating a function by a polynomial of degree k will go to zero a little bit faster than $(x-a)^k$ as x tends to a. But this might not always be the case: it is also possible that increasing the degree of the approximating polynomial does not increase the quality of approximation at all even if the function f to be approximated is infinitely many times differentiable. An example of this behavior is given below, and it is related to the fact that unlike analytic functions, more general functions are not (locally) determined by the values of their derivatives at a single point.

Taylor's theorem is of asymptotic nature: it only tells us that the error R_k in an approximation by a k-th order Taylor polynomial P_k tends to zero faster than any nonzero k-th degree polynomial as $x \to a$. It does not tell us how large the error is in any concrete neighborhood of the center of expansion, but for this purpose there are explicit formulae for the remainder term (given below) which are valid under some additional regularity assumptions on f. These enhanced versions of Taylor's theorem typically lead to uniform estimates for the approximation error in a small neighborhood of the center of expansion, but the estimates do not necessarily hold for neighborhoods which are too large, even if the function f is analytic. In that situation one may have to select several Taylor polynomials with different centers of expansion to have reliable Taylor-approximations of the original function.

There are several things we might do with the remainder term:

1. Estimate the error in using a polynomial $P_k(x)$ of degree k to estimate $f(x)$ on a given interval $(a - r, a + r)$. (The interval and the degree k are fixed; we want to find the error.)

2. Find the smallest degree k for which the polynomial $P_k(x)$ approximates $f(x)$ to within a given error (or tolerance) on a given interval $(a - r, a + r)$. (The interval and the error are fixed; we want to find the degree).

3. Find the largest interval $(a - r, a + r)$ on which $P_k(x)$ approximates $f(x)$ to within a given error ("tolerance"). (The degree and the error are fixed; we want to find the interval).

Taylor's Theorem in one Real Variable

Statement of the Theorem

The precise statement of the most basic version of Taylor's theorem is as follows:

Taylor's theorem. Let $k \geq 1$ be an integer and let the function $f : R \to R$ be k times differentiable at the point $a \in R$. Then there exists a function $h_k : R \to R$ such that,

$$f(x) = f(a) + f'(a)(x-a) + \frac{f''(a)}{2!}(x-a)^2 + \cdots + \frac{f^{(k)}(a)}{k!}(x-a)^k + h_k(x)(x-a)^k,$$

and $\lim_{x \to a} h_k(x) = 0$. This is called the Peano form of the remainder.

The polynomial appearing in Taylor's theorem is the k-th order Taylor polynomial:

$$P_k(x) = f(a) + f'(a)(x-a) + \frac{f''(a)}{2!}(x-a)^2 + \cdots + \frac{f^{(k)}(a)}{k!}(x-a)^k$$

of the function f at the point a. The Taylor polynomial is the unique "asymptotic best fit" polynomial in the sense that if there exists a function $h_k : R \to R$ and a k-th order polynomial p such that:

$$f(x) = p(x) + h_k(x)(x-a)^k, \quad \lim_{x \to a} h_k(x) = 0,$$

then $p = P_k$. Taylor's theorem describes the asymptotic behavior of the remainder term,

$$R_k(x) = f(x) - P_k(x),$$

which is the approximation error when approximating f with its Taylor polynomial. Using the little-o notation, the statement in Taylor's theorem reads as:

$$R_k(x) = o(|x-a|^k), \quad x \to a.$$

Explicit Formulas for the Remainder

Under stronger regularity assumptions on f there are several precise formulae for the remainder term R_k of the Taylor polynomial, the most common ones being the following.

Mean-value forms of the remainder: Let f, $R \to R$ be $k + 1$ times differentiable on the open interval with $f^{(k)}$ continuous on the closed interval between a and x. Then,

$$R_k(x) = \frac{f^{(k+1)}(\xi_L)}{(k+1)!}(x-a)^{k+1}$$

for some real number ξ_L between a and x. This is the Lagrange form of the remainder. Similarly,

$$R_k(x) = \frac{f^{(k+1)}(\xi_C)}{k!}(x-\xi_C)^k(x-a)$$

for some real number ξ_C between a and x. This is the Cauchy form of the remainder.

These refinements of Taylor's theorem are usually proved using the mean value theorem, whence the name. Also other similar expressions can be found. For example, if $G(t)$ is continuous on the closed interval and differentiable with a non-vanishing derivative on the open interval between a and x then,

$$R_k(x) = \frac{f^{(k+1)}(\xi)}{k!}(x-\xi)^k \frac{G(x)-G(a)}{G'(\xi)}$$

for some number ξ between a and x. This version covers the Lagrange and Cauchy forms of the remainder as special cases, and is proved below using Cauchy's mean value theorem.

The statement for the integral form of the remainder is more advanced than the previous ones, and requires understanding of Lebesgue integration theory for the full generality. However, it holds also in the sense of Riemann integral provided the $(k + 1)$th derivative of f is continuous on the closed interval $[a,x]$.

Integral form of the remainder: Let $f^{(k)}$ be absolutely continuous on the closed interval between a and x. Then,

$$R_k(x) = \int_a^x \frac{f^{(k+1)}(t)}{k!}(x-t)^k \, dt.$$

Due to absolute continuity of $f^{(k)}$ on the closed interval between a and x, its derivative $f^{(k+1)}$ exists as an L^1-function, and the result can be proven by a formal calculation using fundamental theorem of calculus and integration by parts.

Estimates for the Remainder

It is often useful in practice to be able to estimate the remainder term appearing in the Taylor approximation, rather than having an exact formula for it. Suppose that f is $(k + 1)$-times continuously differentiable in an interval I containing a. Suppose that there are real constants q and Q such that,

$$q \le f^{(k+1)}(x) \le Q$$

throughout I. Then the remainder term satisfies the inequality,

$$q \frac{(x-a)^{k+1}}{(k+1)!} \le R_k(x) \le Q \frac{(x-a)^{k+1}}{(k+1)!},$$

if $x > a$, and a similar estimate if $x < a$. This is a simple consequence of the Lagrange form of the remainder. In particular if,

$$|f^{(k+1)}(x)| \le M$$

on an interval $I = (a - r, a + r)$ with some $r > 0$, then

$$|R_k(x)| \le M \frac{|x-a|^{k+1}}{(k+1)!} \le M \frac{r^{k+1}}{(k+1)!}$$

for all $x \in (a - r, a + r)$. The second inequality is called a uniform estimate, because it holds uniformly for all x on the interval $(a - r, a + r)$.

Example:

Suppose that we wish to approximate the function $f(x) = e^x$ on the interval $[-1,1]$ while ensuring that the error in the approximation is no more than 10^{-5}. In this example we pretend that we only know the following properties of the exponential function:

$$(*) \qquad e^0 = 1, \qquad \frac{d}{dx} e^x = e^x, \qquad e^x > 0, \qquad x \in \mathbb{R}.$$

From these properties it follows that $f^{(k)}(x) = e^x$ for all k, and in particular, $f^{(k)}(0) = 1$. Hence the k-th order Taylor polynomial of f at 0 and its remainder term in the Lagrange form are given by,

$$P_k(x) = 1 + x + \frac{x^2}{2!} + \cdots + \frac{x^k}{k!}, \qquad R_k(x) = \frac{e^\xi}{(k+1)!} x^{k+1},$$

where ξ is some number between 0 and x. Since e^x is increasing by $(*)$, we can simply use $e^x \le 1$ for $x \in [-1, 0]$ to estimate the remainder on the subinterval $[-1, 0]$. To obtain an upper bound for the remainder on $[0,1]$, we use the property $e^\xi < e^x$ for $0 < \xi < x$ to estimate:

$$e^x = 1 + x + \frac{e^\xi}{2} x^2 < 1 + x + \frac{e^x}{2} x^2, \qquad 0 < x \le 1$$

using the second order Taylor expansion. Then we solve for e^x to deduce that,

$$e^x \le \frac{1+x}{1 - \dfrac{x^2}{2}} = 2 \frac{1+x}{2-x^2} \le 4, \qquad 0 \le x \le 1$$

simply by maximizing the numerator and minimizing the denominator. Combining these estimates for e^x we see that:

$$|R_k(x)| \le \frac{4|x|^{k+1}}{(k+1)!} \le \frac{4}{(k+1)!}, \qquad -1 \le x \le 1,$$

so the required precision is certainly reached when,

$$\frac{4}{(k+1)!} < 10^{-5} \quad \Leftrightarrow \quad 4 \cdot 10^5 < (k+1)! \quad \Leftrightarrow \quad k \ge 9.$$

(Factorial or compute by hand the values 9!=362 880 and 10!=3 628 800.) As a conclusion, Taylor's theorem leads to the approximation:

$$e^x = 1 + x + \frac{x^2}{2!} + \cdots + \frac{x^9}{9!} + R_9(x), \qquad |R_9(x)| < 10^{-5}, \qquad -1 \le x \le 1.$$

For instance, this approximation provides a decimal expression $e \approx 2.71828$, correct up to five decimal places.

Relationship to Analyticity

Taylor Expansions of Real Analytic Functions

Let $I \subset \mathrm{R}$ be an open interval. By definition, a function $f : I \to \mathrm{R}$ is real analytic if it is locally defined by a convergent power series. This means that for every $a \in I$ there exists some $r > 0$ and a sequence of coefficients $c_k \in \mathrm{R}$ such that $(a - r, a + r) \subset I$ and

$$f(x) = \sum_{k=0}^{\infty} c_k (x-a)^k = c_0 + c_1(x-a) + c_2(x-a)^2 + \cdots, \qquad |x-a| < r.$$

In general, the radius of convergence of a power series can be computed from the Cauchy–Hadamard formula,

$$\frac{1}{R} = \limsup_{k \to \infty} |c_k|^{\frac{1}{k}}.$$

This result is based on comparison with a geometric series, and the same method shows that if the power series based on a converges for some $b \in \mathrm{R}$, it must converge uniformly on the closed interval $[a - r_b, a + r_b]$, where $r_b = |b - a|$. Here only the convergence of the power series is considered, and it might well be that $(a - R, a + R)$ extends beyond the domain I of f.

The Taylor polynomials of the real analytic function f at a are simply the finite truncations:

$$P_k(x) = \sum_{j=0}^{k} c_j (x-a)^j, \qquad c_j = \frac{f^{(j)}(a)}{j!}$$

of its locally defining power series, and the corresponding remainder terms are locally given by the analytic functions.

$$R_k(x) = \sum_{j=k+1}^{\infty} c_j(x-a)^j = (x-a)^k h_k(x), \qquad |x-a| < r.$$

Here the functions,

$$\begin{cases} h_k : (a-r, a+r) \to \mathrm{R} \\ h_k(x) = (x-a) \sum_{j=0}^{\infty} c_{k+1+j}(x-a)^j \end{cases}$$

are also analytic, since their defining power series have the same radius of convergence as the original series. Assuming that $[a-r, a+r] \subset I$ and $r < R$, all these series converge uniformly on $(a-r, a+r)$. Naturally, in the case of analytic functions one can estimate the remainder term $R_k(x)$ by the tail of the sequence of the derivatives $f'(a)$ at the center of the expansion, but using complex analysis also another possibility arises.

Taylor's Theorem and Convergence of Taylor Series

The Taylor series of f will converge in some interval, given that all its derivatives are bounded over it and do not grow too fast as k goes to infinity. (However, it is *not always the case* that the Taylor series of f, if it converges, will in fact converge to f, as explained below; f is then said to be non-analytic.)

One might think of the Taylor series,

$$f(x) \approx \sum_{k=0}^{\infty} c_k(x-a)^k = c_0 + c_1(x-a) + c_2(x-a)^2 + \cdots$$

of an infinitely many times differentiable function $f : \mathrm{R} \to \mathrm{R}$ as its "infinite order Taylor polynomial" at a. Now the estimates for the remainder imply that if, for any r, the derivatives of f are known to be bounded over $(a-r, a+r)$, then for any order k and for any $r > 0$ there exists a constant $M_{k,r} > 0$ such that,

$$(*) \quad |R_k(x)| \leqslant M_{k,r} \frac{|x-a|^{k+1}}{(k+1)!}$$

for every $x \in (a-r, a+r)$. Sometimes the constants $M_{k,r}$ can be chosen in such way that $M_{k,r}$ is bounded above, for fixed r and all k. Then the Taylor series of f converges uniformly to some analytic function,

$$\begin{cases} T_f : (a-r, a+r) \to \mathrm{R} \\ T_f(x) = \sum_{k=0}^{\infty} \frac{f^{(k)}(a)}{k!}(x-a)^k \end{cases}$$

(One also gets convergence even if $M_{k,r}$ is not bounded above as long as it grows slowly enough.)

However, even though T_f is always analytic, the case may be that f is not. That is to say, it may well be that an infinitely many times differentiable function f has a Taylor series at a which converges on some open neighborhood of a, but the limit function T_f is different from f. An important example of this phenomenon is provided by the non-analytic smooth function known as a flat function:

$$\begin{cases} f : R \to R \\ f(x) = \begin{cases} e^{-\frac{1}{x^2}} & x > 0 \\ 0 & x \leqslant 0 \end{cases} \end{cases}.$$

Using the chain rule one can show by mathematical induction that for any order k,

$$f^{(k)}(x) = \begin{cases} \dfrac{p_k(x)}{x^{3k}} \cdot e^{-\frac{1}{x^2}} & x > 0 \\ 0 & x \leqslant 0 \end{cases}$$

for some polynomial p_k of degree $2(k - 1)$. The function $e^{-\frac{1}{x^2}}$ tends to zero faster than any polynomial as $x \to 0$, so f is infinitely many times differentiable and $f^{(k)}(0) = 0$ for every positive integer k. Now the estimates for the remainder for the Taylor polynomials show that the Taylor series of f converges uniformly to the zero function on the whole real axis. Nothing is wrong in here:

- The Taylor series of f converges uniformly to the zero function $T_f(x) = 0$.

- The zero function is analytic and every coefficient in its Taylor series is zero.

- The function f is infinitely many times differentiable, but not analytic.

- For any $k \in N$ and $r > 0$ there exists $M_{k,r} > 0$ such that the remainder term for the k-th order Taylor polynomial of f satisfies (*), and is bounded above, for all k and fixed r.

Taylor's Theorem in Complex Analysis

Taylor's theorem generalizes to functions $f : C \to C$ which are complex differentiable in an open subset $U \subset C$ of the complex plane. However, its usefulness is dwarfed by other general theorems in complex analysis. Namely, stronger versions of related results can be deduced for complex differentiable functions $f : U \to C$ using Cauchy's integral formula as follows.

Let $r > 0$ such that the closed disk $B(z, r) \cup S(z, r)$ is contained in U. Then Cauchy's integral formula with a positive parametrization $\gamma(t) = z + re^{it}$ of the circle $S(z, r)$ with $t \in [0, 2\pi]$ gives:

$$f(z) = \frac{1}{2\pi i} \int_\gamma \frac{f(w)}{w - z} dw, \quad f'(z) = \frac{1}{2\pi i} \int_\gamma \frac{f(w)}{(w - z)^2} dw, \quad \ldots, \quad f^{(k)}(z) = \frac{k!}{2\pi i} \int_\gamma \frac{f(w)}{(w - z)^{k+1}} dw.$$

Here all the integrands are continuous on the circle $S(z, r)$, which justifies differentiation under the integral sign. In particular, if f is once complex differentiable on the open set U, then it is actually infinitely many times complex differentiable on U. One also obtains the Cauchy's estimates,

$$|f^{(k)}(z)| \leqslant \frac{k!}{2\pi} \int_\gamma \frac{M_r}{|w - z|^{k+1}} dw = \frac{k! M_r}{r^k}, \quad M_r = \max_{|w - c| = r} |f(w)|$$

for any $z \in U$ and $r > 0$ such that $B(z, r) \cup S(c, r) \subset U$. These estimates imply that the complex Taylor series,

$$T_f(z) = \sum_{k=0}^{\infty} \frac{f^{(k)}(c)}{k!}(z-c)^k$$

of f converges uniformly on any open disk $B(c, r) \subset U$ with $S(c, r) \subset U$ into some function T_f. Furthermore, using the contour integral formulae for the derivatives $f^{(k)}(c)$,

$$T_f(z) = \sum_{k=0}^{\infty} \frac{(z-c)^k}{2\pi i} \int_{\gamma} \frac{f(w)}{(w-c)^{k+1}} dw$$

$$= \frac{1}{2\pi i} \int_{\gamma} \frac{f(w)}{w-c} \sum_{k=0}^{\infty} \left(\frac{z-c}{w-c} \right)^k dw$$

$$= \frac{1}{2\pi i} \int_{\gamma} \frac{f(w)}{w-c} \left(\frac{1}{1 - \dfrac{z-c}{w-c}} \right) dw$$

$$= \frac{1}{2\pi i} \int_{\gamma} \frac{f(w)}{w-c} dw = f(z),$$

so any complex differentiable function f in an open set $U \subset \mathbb{C}$ is in fact complex analytic. All that is said for real analytic functions here holds also for complex analytic functions with the open interval I replaced by an open subset $U \in \mathbb{C}$ and a-centered intervals $(a - r, a + r)$ replaced by c-centered disks $B(c, r)$. In particular, the Taylor expansion holds in the form

$$f(z) = P_k(z) + R_k(z), \quad P_k(z) = \sum_{j=0}^{k} \frac{f^{(j)}(c)}{j!}(z-c)^j,$$

where the remainder term R_k is complex analytic. Methods of complex analysis provide some powerful results regarding Taylor expansions. For example, using Cauchy's integral formula for any positively oriented Jordan curve γ which parametrizes the boundary $\partial W \subset U$ of a region $W \subset U$, one obtains expressions for the derivatives $f^{(j)}(c)$ as above, and modifying slightly the computation for $T_f(z) = f(z)$, one arrives at the exact formula

$$R_k(z) = \sum_{j=k+1}^{\infty} \frac{(z-c)^j}{2\pi i} \int_{\gamma} \frac{f(w)}{(w-c)^{j+1}} dw = \frac{(z-c)^{k+1}}{2\pi i} \int_{\gamma} \frac{f(w) dw}{(w-c)^{k+1}(w-z)}, \quad z \in W.$$

The important feature here is that the quality of the approximation by a Taylor polynomial on the region $W \subset U$ is dominated by the values of the function f itself on the boundary $\partial W \subset U$. Similarly, applying Cauchy's estimates to the series expression for the remainder, one obtains the uniform estimates,

$$|R_k(z)| \leqslant \sum_{j=k+1}^{\infty} \frac{M_r |z-c|^j}{r^j} = \frac{M_r}{r^{k+1}} \frac{|z-c|^{k+1}}{1 - \dfrac{|z-c|}{r}} \leqslant \frac{M_r \beta^{k+1}}{1-\beta}, \quad \frac{|z-c|}{r} \leqslant \beta < 1.$$

Example:

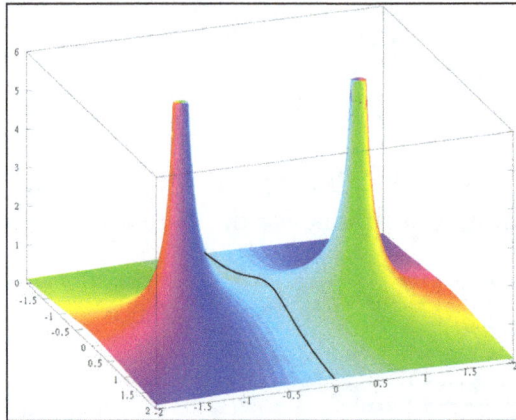

Complex plot of $f(z) = 1/(1 + z^2)$. Modulus is shown by elevation and argument by coloring: cyan=0, blue = $\pi/3$, violet = $2\pi/3$, red = π, yellow=$4\pi/3$, green=$5\pi/3$.

The function

$$\begin{cases} f : \mathrm{R} \to \mathrm{R} \\ f(x) = \dfrac{1}{1+x^2} \end{cases}$$

is real analytic, that is, locally determined by its Taylor series. This function was plotted above to illustrate the fact that some elementary functions cannot be approximated by Taylor polynomials in neighborhoods of the center of expansion which are too large. This kind of behavior is easily understood in the framework of complex analysis. Namely, the function f extends into a meromorphic function

$$\begin{cases} f : \mathrm{C} \cup \{\infty\} \to \mathrm{C} \cup \{\infty\} \\ f(z) = \dfrac{1}{1+z^2} \end{cases}$$

on the compactified complex plane. It has simple poles at $z = i$ and $z = -i$, and it is analytic elsewhere. Now its Taylor series centered at z_0 converges on any disc $B(z_0, r)$ with $r < |z - z_0|$, where the same Taylor series converges at $z \in \mathrm{C}$. Therefore, Taylor series of f centered at 0 converges on $B(0, 1)$ and it does not converge for any $z \in \mathrm{C}$ with $|z| > 1$ due to the poles at i and $-i$. For the same reason the Taylor series of f centered at 1 converges on $B(1, \sqrt{2})$ and does not converge for any $z \in \mathrm{C}$ with $|z - 1| > \sqrt{2}$.

Generalizations of Taylor's Theorem

Higher-order Differentiability

A function $f : \mathrm{R}^n \to \mathrm{R}$ is differentiable at $a \in \mathrm{R}^n$ if and only if there exists a linear functional $L : \mathrm{R}^n \to \mathrm{R}$ and a function $h : \mathrm{R}^n \to \mathrm{R}$ such that,

$$f(\mathrm{x}) = f(\mathrm{a}) + L(\mathrm{x} - \mathrm{a}) + h(\mathrm{x}) \, | \, \mathrm{x} - \mathrm{a} \, |, \qquad \lim_{\mathrm{x} \to \mathbf{a}} h(\mathrm{x}) = 0.$$

If this is the case, then $L = df(a)$ is the (uniquely defined) differential of f at the point a. Furthermore, then the partial derivatives of f exist at a and the differential of f at a is given by,

$$df(\mathbf{a})(\mathbf{v}) = \frac{\partial f}{\partial x_1}(\mathbf{a})v_1 + \cdots + \frac{\partial f}{\partial x_n}(\mathbf{a})v_n.$$

Introduce the multi-index notation

$$|\alpha| = \alpha_1 + \cdots + \alpha_n, \quad \alpha! = \alpha_1! \cdots \alpha_n!, \quad \mathbf{x}^\alpha = x_1^{\alpha_1} \cdots x_n^{\alpha_n}$$

for $\alpha \in \mathbb{N}^n$ and $x \in \mathbb{R}^n$. If all the k-th order partial derivatives of $f : \mathbb{R}^n \to \mathbb{R}$ are continuous at $a \in \mathbb{R}^n$, then by Clairaut's theorem, one can change the order of mixed derivatives at a, so the notation,

$$D^\alpha f = \frac{\partial^{|\alpha|} f}{\partial x_1^{\alpha_1} \cdots \partial x_n^{\alpha_n}}, \qquad |\alpha| \le k$$

for the higher order partial derivatives is justified in this situation. The same is true if all the $(k-1)$-th order partial derivatives of f exist in some neighborhood of a and are differentiable at a. Then we say that f is k times differentiable at the point a.

Taylor's Theorem for Multivariate Functions

Multivariate version of Taylor's theorem. Let $f : \mathbb{R}^n \to \mathbb{R}$ be a k times differentiable function at the point $a \in \mathbb{R}^n$. Then there exists $h_\alpha : \mathbb{R}^n \to \mathbb{R}$ such that

$$f(\mathbf{x}) = \sum_{|\alpha| \le k} \frac{D^\alpha f(\mathbf{a})}{\alpha!}(\mathbf{x} - \mathbf{a})^\alpha + \sum_{|\alpha| = k} h_\alpha(\mathbf{x})(\mathbf{x} - \mathbf{a})^\alpha,$$

and $\quad \lim_{\mathbf{x} \to \mathbf{a}} h_\alpha(\mathbf{x}) = 0.$

If the function $f : \mathbb{R}^n \to \mathbb{R}$ is $k + 1$ times continuously differentiable in the closed ball B, then one can derive an exact formula for the remainder in terms of $(k+1)$-th order partial derivatives of f in this neighborhood. Namely,

$$f(\mathbf{x}) = \sum_{|\alpha| \le k} \frac{D^\alpha f(\mathbf{a})}{\alpha!}(\mathbf{x} - \mathbf{a})^\alpha + \sum_{|\beta| = k+1} R_\beta(\mathbf{x})(\mathbf{x} - \mathbf{a})^\beta,$$

$$R_\beta(\mathbf{x}) = \frac{|\beta|}{\beta!} \int_0^1 (1-t)^{|\beta|-1} D^\beta f\big(\mathbf{a} + t(\mathbf{x} - \mathbf{a})\big) dt.$$

In this case, due to the continuity of $(k+1)$-th order partial derivatives in the compact set B, one immediately obtains the uniform estimates

$$\left| R_\beta(\mathbf{x}) \right| \le \frac{1}{\beta!} \max_{|\alpha| = |\beta|} \max_{\mathbf{y} \in B} |D^\alpha f(\mathbf{y})|, \qquad \mathbf{x} \in B.$$

Example in Two Dimensions

For example, the third-order Taylor polynomial of a smooth function $f: R^2 \rightarrow R$ is, denoting $x - a = v$,

$$P_3(\mathbf{x}) = f(\mathbf{a}) + \frac{\partial f}{\partial x_1}(\mathbf{a})v_1 + \frac{\partial f}{\partial x_2}(\mathbf{a})v_2 + \frac{\partial^2 f}{\partial x_1^2}(\mathbf{a})\frac{v_1^2}{2!} + \frac{\partial^2 f}{\partial x_1 \partial x_2}(\mathbf{a})v_1 v_2 + \frac{\partial^2 f}{\partial x_2^2}(\mathbf{a})\frac{v_2^2}{2!}$$

$$+ \frac{\partial^3 f}{\partial x_1^3}(\mathbf{a})\frac{v_1^3}{3!} + \frac{\partial^3 f}{\partial x_1^2 \partial x_2}(\mathbf{a})\frac{v_1^2 v_2}{2!} + \frac{\partial^3 f}{\partial x_1 \partial x_2^2}(\mathbf{a})\frac{v_1 v_2^2}{2!} + \frac{\partial^3 f}{\partial x_2^3}(\mathbf{a})\frac{v_2^3}{3!}$$

Proofs

Proof for Taylor's Theorem in one Real Variable

Let,

$$h_k(x) = \begin{cases} \dfrac{f(x) - P(x)}{(x-a)^k} & x \neq a \\ 0 & x = a \end{cases}$$

where, as in the statement of Taylor's theorem,

$$P(x) = f(a) + f'(a)(x-a) + \frac{f''(a)}{2!}(x-a)^2 + \cdots + \frac{f^{(k)}(a)}{k!}(x-a)^k.$$

It is sufficient to show that,

$$\lim_{x \to a} h_k(x) = 0.$$

The proof here is based on repeated application of L'Hôpital's rule. Note that, for each $j = 0,1,...,k-1$, $f^{(j)}(a) = P^{(j)}(a)$. Hence each of the first $k-1$ derivatives of the numerator in $h_k(x)$ vanishes at $x = a$, and the same is true of the denominator. Also, since the condition that the function f be k times differentiable at a point requires differentiability up to order $k-1$ in a neighborhood of said point (this is true, because differentiability requires a function to be defined in a whole neighborhood of a point), the numerator and its $k - 2$ derivatives are differentiable in a neighborhood of a. Clearly, the denominator also satisfies said condition, and additionally, doesn't vanish unless $x=a$, therefore all conditions necessary for L'Hopital's rule are fulfilled, and its use is justified. So,

$$\lim_{x \to a} \frac{f(x) - P(x)}{(x-a)^k} = \lim_{x \to a} \frac{\frac{d}{dx}(f(x) - P(x))}{\frac{d}{dx}(x-a)^k} = \cdots = \lim_{x \to a} \frac{\frac{d^{k-1}}{dx^{k-1}}(f(x) - P(x))}{\frac{d^{k-1}}{dx^{k-1}}(x-a)^k}$$

$$= \frac{1}{k!} \lim_{x \to a} \frac{f^{(k-1)}(x) - P^{(k-1)}(x)}{x - a}$$

$$= \frac{1}{k!}(f^{(k)}(a) - f^{(k)}(a)) = 0.$$

where the second to last equality follows by the definition of the derivative at $x = a$.

Derivation for the Mean Value Forms of the Remainder

Let G be any real-valued function, continuous on the closed interval between a and x and differentiable with a non-vanishing derivative on the open interval between a and x and define,

$$F(t) = f(t) + f'(t)(x-t) + \frac{f''(t)}{2!}(x-t)^2 + \cdots + \frac{f^{(k)}(t)}{k!}(x-t)^k.$$

For $t \in [a,x]$. Then, by Cauchy's mean value theorem,

$$(*) \quad \frac{F'(\ )}{G'(\ )} \quad \frac{F(x)}{G(x)} \quad \frac{F(a)}{G(a)}$$

for some ξ on the open interval between a and x. Note that here the numerator $F(x) - F(a) = R_k(x)$ is exactly the remainder of the Taylor polynomial for $f(x)$. Compute,

$$F'(t) = f'(t) + \left(f''(t)(x-t) - f'(t)\right) + \left(\frac{f^{(3)}(t)}{2!}(x-t)^2 - \frac{f^{(2)}(t)}{1!}(x-t)\right) + \cdots$$

$$\cdots + \left(\frac{f^{(k+1)}(t)}{k!}(x-t)^k - \frac{f^{(k)}(t)}{(k-1)!}(x-t)^{k-1}\right) = \frac{f^{(k+1)}(t)}{k!}(x-t)^k,$$

plug it into (*) and rearrange terms to find that,

$$R_k(x) = \frac{f^{(k+1)}(\xi)}{k!}(x-\xi)^k \frac{G(x) - G(a)}{G'(\xi)}.$$

This is the form of the remainder term mentioned after the actual statement of Taylor's theorem with remainder in the mean value form. The Lagrange form of the remainder is found by choosing $G(t) = (t-x)^{k+1}$ and the Cauchy form by choosing $G(t) = t - a$.

Remark. Using this method one can also recover the integral form of the remainder by choosing,

$$G(t) = \int_a^t \frac{f^{(k+1)}(s)}{k!}(x-s)^k \, ds,$$

but the requirements for f needed for the use of mean value theorem are too strong, if one aims to prove the claim in the case that $f^{(k)}$ is only absolutely continuous. However, if one uses Riemann integral instead of Lebesgue integral, the assumptions cannot be weakened.

Derivation for the Integral form of the Remainder

Due to absolute continuity of $f^{(k)}$ on the closed interval between a and x its derivative $f^{(k+1)}$ exists as an L^1-function, and we can use fundamental theorem of calculus and integration by parts. This same proof applies for the Riemann integral assuming that $f^{(k)}$ is continuous on the closed interval and differentiable on the open interval between a and x, and this leads to the same result than using the mean value theorem.

The fundamental theorem of calculus states that,

$$f(x) = f(a) + \int_a^x f'(t)dt.$$

Now we can integrate by parts and use the fundamental theorem of calculus again to see that,

$$f(x) = f(a) + \left(xf'(x) - af'(a)\right) - \int_a^x tf''(t)dt$$

$$= f(a) + x\left(f'(a) + \int_a^x f''(t)dt\right) - af'(a) - \int_a^x tf''(t)dt$$

$$= f(a) + (x-a)f'(a) + \int_a^x (x-t)f''(t)dt,$$

which is exactly Taylor's theorem with remainder in the integral form in the case $k=1$. The general statement is proved using induction. Suppose that,

$$(*) \quad f(x) = f(a) + \frac{f'(a)}{1!}(x-a) + \cdots + \frac{f^{(k)}(a)}{k!}(x-a)^k + \int_a^x \frac{f^{(k+1)}(t)}{k!}(x-t)^k\, dt.$$

Integrating the remainder term by parts we arrive at,

$$\int_a^x \frac{f^{(k+1)}(t)}{k!}(x-t)^k\, dt = -\left[\frac{f^{(k+1)}(t)}{(k+1)k!}(x-t)^{k+1}\right]_a^x + \int_a^x \frac{f^{(k+2)}(t)}{(k+1)k!}(x-t)^{k+1}\, dt$$

$$= \frac{f^{(k+1)}(a)}{(k+1)!}(x-a)^{k+1} + \int_a^x \frac{f^{(k+2)}(t)}{(k+1)!}(x-t)^{k+1}\, dt.$$

Substituting this into the formula in (*) shows that if it holds for the value k, it must also hold for the value $k + 1$. Therefore, since it holds for $k = 1$, it must hold for every positive integer k.

Derivation for the Cauchy form of the Remainder

To the integral form of the remainder, we can apply the mean value theorem for integral.

$$\int_a^x \frac{f^{(k+1)}(t)}{k!}(x-t)^k\, dt = \frac{f^{(k+1)}(\xi)}{k!}(x-\xi)^k \int_a^x dt$$

$$= \frac{f^{(k+1)}(\xi)}{k!}(x-\xi)^k(x-a),$$

where $\xi \in \{a + \theta(x-a) : 0 < \theta < 1\}$

So, The Cauchy form of the remainder is hold.

Derivation for the Remainder of Multivariate Taylor Polynomials

We prove the special case, where $f \colon \mathrm{R}^n \to \mathrm{R}$ has continuous partial derivatives up to the order $k+1$ in some closed ball B with center a. The strategy of the proof is to apply the one-variable case of Taylor's theorem to the restriction of f to the line segment adjoining x and a. Parametrize the line segment between a and x by $u(t) = a + t(x - a)$. We apply the one-variable version of Taylor's theorem to the function $g(t) = f(u(t))$:

$$f(\mathbf{x}) = g(1) = g(0) + \sum_{j=1}^{k} \frac{1}{j!} g^{(j)}(0) + \int_0^1 \frac{(1-t)^k}{k!} g^{(k+1)}(t)\,dt.$$

Applying the chain rule for several variables gives:

$$g^{(j)}(t) = \frac{d^j}{dt^j} f(u(t)) = \frac{d^j}{dt^j} f(\mathbf{a} + t(\mathbf{x} - \mathbf{a}))$$

$$= \sum_{|\alpha|=j} \binom{j}{\alpha} (D^\alpha f)(\mathbf{a} + t(\mathbf{x} - \mathbf{a}))(\mathbf{x} - \mathbf{a})^\alpha$$

where $\binom{j}{\alpha}$ is the multinomial coefficient. Since $\dfrac{1}{j!}\dbinom{j}{\alpha} = \dfrac{1}{\alpha!}$, we get

$$f(\mathbf{x}) = f(\mathbf{a}) + \sum_{1 \le |\alpha| \le k} \frac{1}{\alpha!} (D^\alpha f)(\mathbf{a})(\mathbf{x} - \mathbf{a})^\alpha + \sum_{|\alpha|=k+1} \frac{k+1}{\alpha!} (\mathbf{x} - \mathbf{a})^\alpha \int_0^1 (1-t)^k (D^\alpha f)(\mathbf{a} + t(\mathbf{x} - \mathbf{a}))\,dt.$$

STOKES' THEOREM

In vector calculus, and more generally differential geometry, Stokes' theorem (sometimes spelled Stokes's theorem, and also called the generalized Stokes theorem or the Stokes–Cartan theorem) is a statement about the integration of differential forms on manifolds, which both simplifies and generalizes several theorems from vector calculus. Stokes' theorem says that the integral of a differential form ω over the boundary of some orientable manifold Ω is equal to the integral of its exterior derivative $d\omega$ over the whole of Ω, i.e.,

$$\int_{\partial\Omega} \omega = \int_\Omega d\omega.$$

Stokes' theorem was formulated in its modern form by Élie Cartan in 1945, following earlier work on the generalization of the theorems of vector calculus by Vito Volterra, Édouard Goursat, and Henri Poincaré.

This modern form of Stokes' theorem is a vast generalization of a classical result that Lord Kelvin communicated to George Stokes in a letter dated July 2, 1850. Stokes set the theorem as a question on the 1854 Smith's Prize exam, which led to the result bearing his name. It was first published by Hermann Hankel in 1861. This classical Kelvin–Stokes theorem relates the surface integral of the curl of a vector field F over a surface in Euclidean three-space to the line integral of the vector field over its boundary: Let γ: $[a, b] \to \mathbb{R}^2$ be a piecewise smooth Jordan plane curve. The Jordan curve theorem implies that γ divides $\mathbb{R}^2$ into two components, a compact one and another that is non-compact. Let D denote the compact part that is bounded by γ and suppose ψ: $D \to \mathbb{R}^3$ is smooth, with $S := \psi(D)$. If Γ is the space curve defined by $\Gamma(t) = \psi(\gamma(t))$ and F is a smooth vector field on $\mathbb{R}^3$, then:

$$\oint_\Gamma F \cdot d\Gamma = \iint_S \nabla \times F \cdot dS$$

This classical statement, along with the classical divergence theorem, the fundamental theorem of calculus, and Green's theorem are simply special cases of the general formulation stated above.

The fundamental theorem of calculus states that the integral of a function f over the interval $[a, b]$ can be calculated by finding an antiderivative F of f:

$$\int_a^b f(x)dx = F(b) - F(a).$$

Stokes' theorem is a vast generalization of this theorem in the following sense.

- By the choice of F, $dF/dx = f(x)$. In the parlance of differential forms, this is saying that $f(x)\,dx$ is the exterior derivative of the 0-form, i.e. function, F: in other words, that $dF = f\,dx$. The general Stokes theorem applies to higher differential forms ω instead of just 0-forms such as F.

- A closed interval $[a, b]$ is a simple example of a one-dimensional manifold with boundary. Its boundary is the set consisting of the two points a and b. Integrating f over the interval may be generalized to integrating forms on a higher-dimensional manifold. Two technical conditions are needed: the manifold has to be orientable, and the form has to be compactly supported in order to give a well-defined integral.

- The two points a and b form the boundary of the closed interval. More generally, Stokes' theorem applies to oriented manifolds M with boundary. The boundary ∂M of M is itself a manifold and inherits a natural orientation from that of M. For example, the natural orientation of the interval gives an orientation of the two boundary points. Intuitively, a inherits the opposite orientation as b, as they are at opposite ends of the interval. So, "integrating" F over two boundary points a, b is taking the difference $F(b) - F(a)$.

In even simpler terms, one can consider the points as boundaries of curves, that is as 0-dimensional boundaries of 1-dimensional manifolds. So, just as one can find the value of an integral ($f\,dx = dF$) over a 1-dimensional manifold ($[a, b]$) by considering the anti-derivative (F) at the 0-dimensional boundaries ($\{a, b\}$), one can generalize the fundamental theorem of calculus, with a few additional caveats, to deal with the value of integrals ($d\omega$) over n-dimensional manifolds (Ω) by considering the antiderivative (ω) at the ($n - 1$)-dimensional boundaries ($\partial\Omega$) of the manifold.

So the fundamental theorem reads:

$$\int_{[a,b]} f(x)dx = \int_{[a,b]} dF = \int_{\{a\}^- \cup \{b\}^+} F = F(b) - F(a).$$

Formulation for Smooth Manifolds with Boundary

Let Ω be an oriented smooth manifold with boundary of dimension n and let α be a smooth n-differential form that is compactly supported on Ω. First, suppose that α is compactly supported in the domain of a single, oriented coordinate chart $\{U, \varphi\}$. In this case, we define the integral of α over Ω as,

$$\int_\Omega \alpha = \int_{\varphi(U)} \left(\varphi^{-1}\right)^* \alpha,$$

i.e., via the pullback of α to R^n.

More generally, the integral of α over Ω is defined as follows: Let $\{\psi_i\}$ be a partition of unity associated with a locally finite cover $\{U_i, \varphi_i\}$ of (consistently oriented) coordinate charts, then define the integral,

$$\int_\Omega \alpha \equiv \sum_i \int_{U_i} \psi_i \alpha,$$

where each term in the sum is evaluated by pulling back to R^n as described above. This quantity is well-defined; that is, it does not depend on the choice of the coordinate charts, nor the partition of unity.

The generalized Stokes theorem reads:

Theorem: (Stokes–Cartan) If ω is a smooth $(n-1)$-form with compact support on smooth n-dimensional manifold-with-boundary Ω, $\partial\Omega$ denotes the boundary of Ω given the induced orientation, and $i : \partial\Omega \mapsto \Omega$ is the inclusion map, then

$$\int_\Omega d\omega = \int_{\partial\Omega} i^*\omega.$$

Conventionally, $\int_{\partial\Omega} i^*\omega$ is abbreviated as $\int_{\partial\Omega} \omega$, since the pullback of a differential form by the inclusion map is simply its restriction to its domain: $i^*\omega = \omega|_{\partial\Omega}$. Here d is the exterior derivative, which is defined using the manifold structure only. The right-hand side is sometimes written as $\oint_{\partial\Omega} \omega$. to stress the fact that the $(n-1)$-manifold $\partial\Omega$ has no boundary. (This fact is also an implication of Stokes' theorem, since for a given smooth n-dimensional manifold Ω, application of the theorem twice gives $\int_{\partial(\partial\Omega)} \omega = \int_\Omega d(d\omega) = 0$ for any $(n-2)$-form ω, which implies that $\partial(\partial\Omega) = \theta$.)

The right-hand side of the equation is often used to formulate *integral* laws; the left-hand side then leads to equivalent *differential* formulations .

The theorem is often used in situations where Ω is an embedded oriented submanifold of some bigger manifold, often R^k, on which the form ω is defined.

Topological Preliminaries; Integration Over Chains

Let M be a smooth manifold. A (smooth) singular k-simplex in M is defined as a smooth map from the standard simplex in R^k to M. The group $C_k(M, Z)$ of singular k-chains on M is defined to be the free abelian group on the set of singular k-simplices in M. These groups, together with the boundary map, ∂, define a chain complex. The corresponding homology (resp. cohomology) group is isomorphic to the usual singular homology group $H_k(M, Z)$ (resp. the singular cohomology group $H^k(M, Z)$), defined using continuous rather than smooth simplices in M.

On the other hand, the differential forms, with exterior derivative, d, as the connecting map, form a cochain complex, which defines the de Rham cohomology groups $H^k{}_{dR}(M, R)$.

Differential k-forms can be integrated over a k-simplex in a natural way, by pulling back to R^k. Extending by linearity allows one to integrate over chains. This gives a linear map from the space of k-forms to the kth group of singular cochains, $C^k(M, Z)$, the linear functionals on $C_k(M, Z)$. In other words, a k-form ω defines a functional,

$$I(\omega)(c) = \oint_c \omega$$

on the k-chains. Stokes' theorem says that this is a chain map from de Rham cohomology to singular cohomology with real coefficients; the exterior derivative, d, behaves like the *dual* of ∂ on forms. This gives a homomorphism from de Rham cohomology to singular cohomology. On the level of forms, this means:

1. Closed forms, i.e., $d\omega = 0$, have zero integral over *boundaries*, i.e. over manifolds that can be written as $\partial \sum_c M_c$, and

2. Exact forms, i.e., $\omega = d\sigma$, have zero integral over *cycles*, i.e. if the boundaries sum up to the empty set: $\sum_c M_c = \emptyset$.

De Rham's theorem shows that this homomorphism is in fact an isomorphism. So the converse to 1 and 2 above hold true. In other words, if $\{c_i\}$ are cycles generating the kth homology group, then for any corresponding real numbers, $\{a_i\}$, there exist a closed form, ω, such that

$$\oint_{c_i} \omega = a_i \, ,$$

and this form is unique up to exact forms.

Stokes' theorem on smooth manifolds can be derived from Stokes' theorem for chains in smooth manifolds, and vice versa. Formally stated, the latter reads:

Theorem: (Stokes' theorem for chains) If c is a smooth k-chain in a smooth manifold M, and ω is a smooth $(k-1)$-form on M, then

$$\int_{\partial c} \omega = \int_c d\omega .$$

Underlying Principle

To simplify these topological arguments, it is worthwhile to examine the underlying principle by considering an example for $d = 2$ dimensions. The essential idea can be understood by the diagram on the left, which shows that, in an oriented tiling of a manifold, the interior paths are traversed in opposite directions; their contributions to the path integral thus cancel each other pairwise. As a consequence, only the contribution from the boundary remains. It thus suffices to prove Stokes' theorem for sufficiently fine tilings (or, equivalently, simplices), which usually is not difficult.

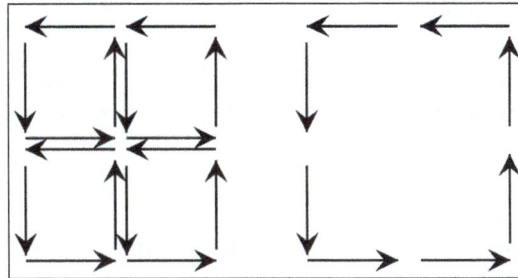

Generalization to Rough Sets

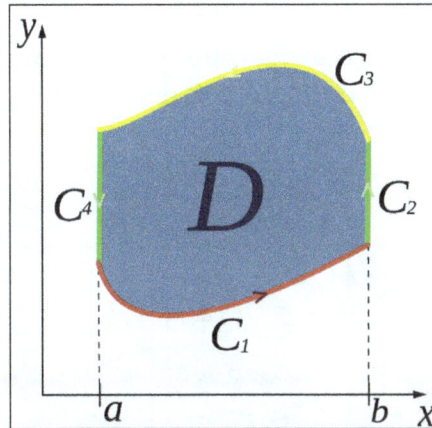

A region (here called D instead of Ω) with piecewise smooth boundary. This is a manifold with corners, so its boundary is not a smooth manifold.

The formulation above, in which Ω is a smooth manifold with boundary, does not suffice in many applications. For example, if the domain of integration is defined as the plane region between two x-coordinates and the graphs of two functions, it will often happen that the domain has corners. In such a case, the corner points mean that Ω is not a smooth manifold with boundary, and so the statement of Stokes' theorem given above does not apply. Nevertheless, it is possible to check that the conclusion of Stokes' theorem is still true. This is because Ω and its boundary are well-behaved away from a small set of points (a measure zero set).

A version of Stokes' theorem that extends to rough domains was proved by Whitney. Assume that D is a connected bounded open subset of $\mathbb{R}^n$. Call D a *standard domain* if it satisfies the following property: There exists a subset P of ∂D, open in ∂D, whose complement in ∂D has Hausdorff $(n - 1)$-measure zero; and such that every point of P has a *generalized normal vector*. This is a vector $v(x)$ such that, if a coordinate system is chosen so that $v(x)$ is the first basis vector, then, in an open neighborhood around x, there exists a smooth function $f(x_2, ..., x_n)$ such that P is the graph $\{ x_1 = f(x_2, ..., x_n) \}$ and D is the region $\{ x_1 < f(x_2, ..., x_n) \}$. Whitney remarks that the boundary of a standard domain is the union of a set of zero Hausdorff $(n - 1)$-measure and a finite or countable union of smooth $(n - 1)$-manifolds, each of which has the domain on only one side. He then proves that if D is a standard domain in $\mathbb{R}^n$, ω is an $(n-1)$-form which is defined, continuous, and bounded on $D \cup P$, smooth on D, integrable on P, and such that $d\omega$ is integrable on D, then Stokes' theorem holds, that is,

$$\int_P \omega = \int_D d\omega$$

The study of measure-theoretic properties of rough sets leads to geometric measure theory. Even more general versions of Stokes' theorem have been proved by Federer and by Harrison.

Special Cases

The general form of the Stokes theorem using differential forms is more powerful and easier to use than the special cases. The traditional versions can be formulated using Cartesian coordinates without the machinery of differential geometry, and thus are more accessible. Further, they are older and their names are more familiar as a result. The traditional forms are often considered more convenient by practicing scientists and engineers but the non-naturalness of the traditional formulation becomes apparent when using other coordinate systems, even familiar ones like spherical or cylindrical coordinates. There is potential for confusion in the way names are applied, and the use of dual formulations.

Kelvin–Stokes Theorem

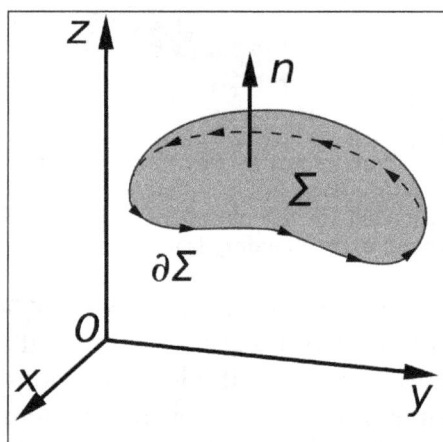

An illustration of the Kelvin–Stokes theorem, with surface Σ, its boundary $\partial\Sigma$ and the "normal" vector n.

This is a (dualized) (1 + 1)-dimensional case, for a 1-form (dualized because it is a statement about vector fields). This special case is often just referred to as *Stokes' theorem* in many introductory university vector calculus courses and is used in physics and engineering. It is also sometimes known as the curl theorem.

The classical Kelvin–Stokes theorem relates the surface integral of the curl of a vector field over a surface Σ in Euclidean three-space to the line integral of the vector field over its boundary. It is a special case of the general Stokes theorem (with $n = 2$) once we identify a vector field with a 1-form using the metric on Euclidean 3-space. The curve of the line integral, $\partial\Sigma$, must have positive orientation, meaning that $\partial\Sigma$ points counterclockwise when the surface normal, n, points toward the viewer.

One consequence of the Kelvin–Stokes theorem is that the field lines of a vector field with zero curl cannot be closed contours. The formula can be rewritten as:

Suppose $F=(P(x,y,z),Q(x,y,z),R(x,y,z))$ is defined in region with smooth surface Σ and has first-order continuous partial derivatives. Then,

$$\iint_\Sigma \left(\left(\frac{\partial R}{\partial y} - \frac{\partial Q}{\partial z} \right) dy\,dz + \left(\frac{\partial P}{\partial z} - \frac{\partial R}{\partial x} \right) dz\,dx + \left(\frac{\partial Q}{\partial x} - \frac{\partial P}{\partial y} \right) dx\,dy \right)$$
$$= \oint_{\partial\Sigma} \left(P\,dx + Q\,dy + R\,dz \right),$$

where P, Q and R are the components of F, and $\partial\Sigma$ is boundary of region with smooth surface Σ.

Green's Theorem

Green's theorem is immediately recognizable as the third integrand of both sides in the integral in terms of P, Q, and R cited above.

In Electromagnetism

Two of the four Maxwell equations involve curls of 3-D vector fields, and their differential and integral forms are related by the Kelvin–Stokes theorem. Caution must be taken to avoid cases with moving boundaries: the partial time derivatives are intended to exclude such cases. If moving boundaries are included, interchange of integration and differentiation introduces terms related to boundary motion not included in the results below:

Name	Differential form	Integral form (using Kelvin–Stokes theorem plus relativistic invariance, $\int \partial/\partial t \,...\,\to d/dt \int \,...$)
Maxwell–Faraday equation Faraday's law of induction:	$\nabla \times E = -\dfrac{\partial B}{\partial t}$	$\oint_C E \cdot d\mathrm{l} = \iint_S \nabla \times E \cdot dA = -\iint_S \dfrac{\partial B}{\partial t} \cdot dA$ (with C and S not necessarily stationary)
Ampère's law (with Maxwell's extension):	$\nabla \times H = J + \dfrac{\partial D}{\partial t}$	$\oint_C H \cdot d\mathrm{l} = \iint_S \nabla \times H \cdot dA = \iint_S J \cdot dA + \iint_S \dfrac{\partial D}{\partial t} \cdot dA$ (with C and S not necessarily stationary)

The above listed subset of Maxwell's equations are valid for electromagnetic fields expressed in SI units. In other systems of units, such as CGS or Gaussian units, the scaling factors for the terms differ. For example, in Gaussian units, Faraday's law of induction and Ampère's law take the forms:

$$\nabla \times E = -\frac{1}{c}\frac{\partial B}{\partial t},$$
$$\nabla \times H = \frac{1}{c}\frac{\partial D}{\partial t} + \frac{4\pi}{c}J,$$

respectively, where c is the speed of light in vacuum.

DIVERGENCE THEOREM

In vector calculus, the divergence theorem, also known as Gauss's theorem or Ostrogradsky's theorem, is a result that relates the flow (that is, flux) of a vector field through a surface to the behavior of the tensor field inside the surface.

More precisely, the divergence theorem states that the outward flux of a tensor field through a closed surface is equal to the volume integral of the divergence over the region inside the surface. Intuitively, it states that *the sum of all sources (with sinks regarded as negative sources) gives the net flux out of a region.*

The divergence theorem is an important result for the mathematics of physics and engineering, in particular in electrostatics and fluid dynamics.

In physics and engineering, the divergence theorem is usually applied in three dimensions. However, it generalizes to any number of dimensions. In one dimension, it is equivalent to the fundamental theorem of calculus. In two dimensions, it is equivalent to Green's theorem.

If a fluid is flowing in some area, then the rate at which fluid flows out of a certain region within that area can be calculated by adding up the sources inside the region and subtracting the sinks. The fluid flow is represented by a first order tensor (or vector) field, and the vector field's divergence at a given point describes the strength of the source or sink there. So, integrating the field's divergence over the interior of the region should equal the integral of the vector field over the region's boundary. The divergence theorem says that this is true.

The divergence theorem is employed in any conservation law which states that the volume total of all sinks and sources, that is the volume integral of the divergence, is equal to the net flow across the volume's boundary.

Mathematical Statement

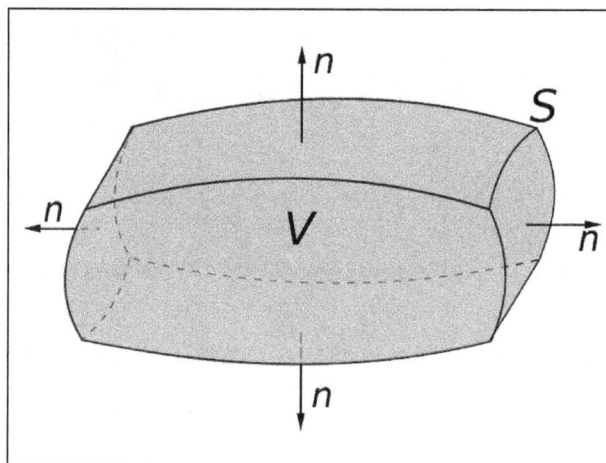

A region V bounded by the surface $S = \partial V$ with the surface normal n

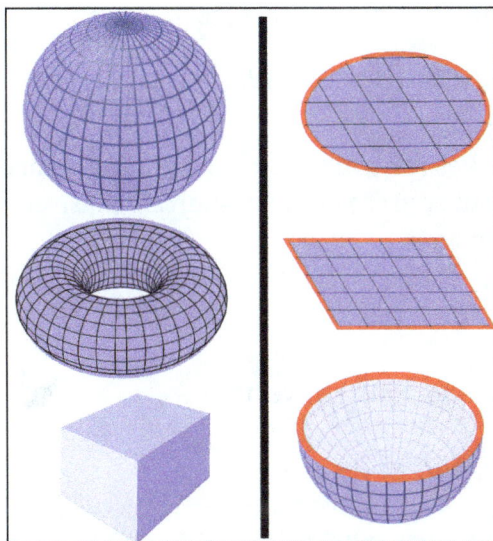

The divergence theorem can be used to calculate a flux through a closed surface that fully encloses a volume, like any of the surfaces on the left. It can *not* directly be used to calculate the flux through surfaces with boundaries, like those on the right. (Surfaces are blue, boundaries are red.)

Suppose V is a subset of $\mathbb{R}^n$ (in the case of $n = 3$, V represents a volume in three-dimensional space) which is compact and has a piecewise smooth boundary S (also indicated with $\partial V = S$). If F is a continuously differentiable vector field defined on a neighborhood of V, then we have:

$$\iiint_V (\nabla \cdot F)\, dV = \oiint_S (F \cdot n)\, dS.$$

The left side is a volume integral over the volume V, the right side is the surface integral over the boundary of the volume V. The closed manifold ∂V is quite generally the boundary of V oriented by outward-pointing normals, and n is the outward pointing unit normal field of the boundary ∂V. (dS may be used as a shorthand for $n dS$.) The symbol within the two integrals stresses once more that ∂V is a *closed* surface. In terms of the intuitive description above, the left-hand side of the equation represents the total of the sources in the volume V, and the right-hand side represents the total flow across the boundary S.

Corollaries

By replacing F in the divergence theorem with specific forms, other useful identities can be derived.

- With $F \rightarrow F g$ for a scalar function g and a vector field F,

$$\iiint_V \left[F \cdot (\nabla g) + g (\nabla \cdot F) \right] dV = \oiint_S g\, F \cdot n\, dS.$$

  A special case of this is $F = \nabla f$, in which case the theorem is the basis for Green's identities.

- With $F \rightarrow F \times G$ for two vector fields F and G,

$$\iiint_V \left[G \cdot (\nabla \times F) - F \cdot (\nabla \times G) \right] dV = \oiint_S (F \times G) \cdot n\, dS.$$

With $F \to f\mathbf{c}$ for a scalar function f and vector field $\mathbf{c}$:

$$\iiint_V \mathbf{c} \cdot \nabla f \, dV = \oiint_S (\mathbf{c}f) \cdot \mathbf{n} \, dS - \iiint_V f(\nabla \cdot \mathbf{c}) dV.$$

The last term on the right vanishes for constant $\mathbf{c}$ or any divergence free (solenoidal) vector field, e.g. Incompressible flows without sources or sinks such as phase change or chemical reactions etc. In particular, taking $\mathbf{c}$ to be constant:

$$\iiint_V \nabla f \, dV = \oiint_S f \, d\mathbf{S}.$$

With $F \to \mathbf{c} \times F$ for vector field F and constant vector $\mathbf{c}$:

$$\iiint_V \mathbf{c} \cdot (\nabla \times F) dV = \oiint_S (F \times \mathbf{c}) \cdot \mathbf{n} \, dS.$$

By reordering the triple product on the right hand side and taking out the constant vector of the integral,

$$\iiint_V (\nabla \times F) dV \cdot \mathbf{c} = \oiint_S (d\mathbf{S} \times F) \cdot \mathbf{c}.$$

Hence,

$$\iiint_V (\nabla \times F) dV = \oiint_S \mathbf{n} \times F \, dS.$$

Example:

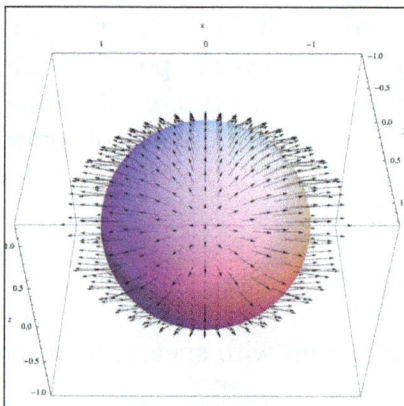

The vector field corresponding to the example shown. Vectors may point into or out of the sphere.

Suppose we wish to evaluate,

$$\oiint_S F \cdot \mathbf{n} \, dS,$$

where S is the unit sphere defined by,

$$S = \left\{ (x, y, z) \in \mathbb{R}^3 : x^2 + y^2 + z^2 = 1 \right\},$$

and F is the vector field,

$$F = 2x\mathbf{i} + y^2\mathbf{j} + z^2\mathbf{k}.$$

The direct computation of this integral is quite difficult, but we can simplify the derivation of the result using the divergence theorem, because the divergence theorem says that the integral is equal to:

$$\iiint_W (\nabla \cdot F)dV = 2\iiint_W (1 + y + z)dV = 2\iiint_W dV + 2\iiint_W y\,dV + 2\iiint_W z\,dV,$$

where W is the unit ball:

$$W = \left\{(x, y, z) \in \mathbb{R}^3 : x^2 + y^2 + z^2 \leq 1\right\}.$$

Since the function y is positive in one hemisphere of W and negative in the other, in an equal and opposite way, its total integral over W is zero. The same is true for z:

$$\iiint_W y\,dV = \iiint_W z\,dV = 0.$$

Therefore,

$$\oiint_S F \cdot \mathbf{n}\,dS = 2\iiint_W dV = \frac{8\pi}{3},$$

because the unit ball W has volume $4\pi/3$.

Applications

Differential form and Integral form of Physical Laws

As a result of the divergence theorem, a host of physical laws can be written in both a differential form (where one quantity is the divergence of another) and an integral form (where the flux of one quantity through a closed surface is equal to another quantity). Three examples are Gauss's law (in electrostatics), Gauss's law for magnetism, and Gauss's law for gravity.

Continuity Equations

Continuity equations offer more examples of laws with both differential and integral forms, related to each other by the divergence theorem. In fluid dynamics, electromagnetism, quantum mechanics, relativity theory, and a number of other fields, there are continuity equations that describe the conservation of mass, momentum, energy, probability, or other quantities. Generically, these equations state that the divergence of the flow of the conserved quantity is equal to the distribution of *sources* or *sinks* of that quantity. The divergence theorem states that any such continuity equation can be written in a differential form (in terms of a divergence) and an integral form (in terms of a flux).

Inverse-square Laws

Any *inverse-square law* can instead be written in a *Gauss's law*-type form (with a differential and integral form, as described above). Two examples are Gauss's law (in electrostatics), which follows from the inverse-square Coulomb's law, and Gauss's law for gravity, which follows from the inverse-square Newton's law of universal gravitation. The derivation of the Gauss's law-type equation from the inverse-square formulation or vice versa is exactly the same in both cases.

Examples:

To verify the planar variant of the divergence theorem for a region R:

$$R = \left\{(x, y) \in \mathbb{R}^2 : x^2 + y^2 \le 1\right\},$$

and the vector field:

$$F(x, y) = 2y\mathbf{i} + 5x\mathbf{j}.$$

The boundary of R is the unit circle, C, that can be represented parametrically by:

$$x = \cos(s), \quad y = \sin(s)$$

such that $0 \le s \le 2\pi$ where s units is the length arc from the point $s = 0$ to the point P on C. Then a vector equation of C is

$$C(s) = \cos(s)\mathbf{i} + \sin(s)\mathbf{j}.$$

At a point P on C:

$$P = (\cos(s), \sin(s)) \Rightarrow F = 2\sin(s)\mathbf{i} + 5\cos(s)\mathbf{j}.$$

Therefore,

$$\oint_C F \cdot n \, ds = \int_0^{2\pi} (2\sin(s)\mathbf{i} + 5\cos(s)\mathbf{j}) \cdot (\cos(s)\mathbf{i} + \sin(s)\mathbf{j}) \, ds$$

$$= \int_0^{2\pi} (2\sin(s)\cos(s) + 5\sin(s)\cos(s)) \, ds$$

$$= 7\int_0^{2\pi} \sin(s)\cos(s) \, ds$$

$$= 0.$$

Because $M = 2y$, $\partial M/\partial x = 0$, and because $N = 5x$, $\partial N/\partial y = 0$. Thus

$$\iint_R \nabla \cdot F \, dA = \iint_R \left(\frac{\partial M}{\partial x} + \frac{\partial N}{\partial y}\right) dA = 0.$$

Applied Example

Let's say we wanted to evaluate the flux of the following vector field defined by $F = 2x^2 + 2y^2 + 2z^2$ bounded by the following inequalities:

$$\{0 \le x \le 3\}\{-2 \le y \le 2\}\{0 \le z \le 2\pi\}$$

We know from the Divergence Theorem that: $\iiint_V (\nabla \cdot F) dV = \oiint_S (F \cdot n) dS$.

We need to determine $\nabla \cdot F$

The divergence of a three dimensional vector field, F, is defined as $\dfrac{\partial F_x}{\partial x} + \dfrac{\partial F_y}{\partial y} + \dfrac{\partial F_z}{\partial z}$

Thus, we can set up the following integrals:

$$\oiint_S F \cdot n \, dS$$

$$= \iiint_V \nabla \cdot F \, dV$$

$$= \iiint_V \frac{\partial F_x}{\partial x} + \frac{\partial F_y}{\partial y} + \frac{\partial F_z}{\partial z} dV$$

$$= \iiint_V 4x + 4y + 4z \, dV$$

$$= \int_0^3 \int_{-2}^2 \int_0^{2\pi} 4x + 4y + 4z \, dV$$

Now that we have set up the integral, we can evaluate it.

$$\int_0^3 \int_{-2}^2 \int_0^{2\pi} 4x + 4y + 4z \, dV$$

$$= \int_{-2}^2 \int_0^{2\pi} 12y + 12z + 18 \, dy \, dz$$

$$= \int_0^{2\pi} 24(2z + 3) \, dz$$

$$= 48\pi \cdot (2\pi + 3)$$

$$\approx 1399.87134$$

Generalizations

Multiple Dimensions

One can use the general Stokes' Theorem to equate the n-dimensional volume integral of the divergence of a vector field F over a region U to the ($n - 1$)-dimensional surface integral of F over the boundary of U:

$$\underbrace{\int \ldots \int_U \nabla . F \, dV}_{n} = \underbrace{\oint \ldots \oint_{\partial U} F . n \, dS}_{n-1}$$

This equation is also known as the divergence theorem.

When $n = 2$, this is equivalent to Green's theorem.

When $n = 1$, it reduces to the Fundamental theorem of calculus.

Tensor Fields

Writing the theorem in Einstein notation:

$$\iiint_V \frac{\partial F_i}{\partial x_i} dV = \oiint_s F_i \, n_i \, dS$$

suggestively, replacing the vector field F with a rank-n tensor field T, this can be generalized to:

$$\iiint_V \frac{\partial T_{i_1 i_2 \cdots i_q \cdots i_n}}{\partial x_{i_q}} dV = \oiint_s T_{i_1 i_2 \cdots i_q \cdots i_n} n_{i_q} \, dS.$$

where on each side, tensor contraction occurs for at least one index. This form of the theorem is still in 3d, each index takes values 1, 2, and 3. It can be generalized further still to higher (or lower) dimensions (for example to 4d spacetime in general relativity).

GRADIENT THEOREM

The gradient theorem, also known as the fundamental theorem of calculus for line integrals, says that a line integral through a gradient field can be evaluated by evaluating the original scalar field at the endpoints of the curve.

Let $\varphi : U \subseteq \mathbb{R}^n \to \mathbb{R}$ and γ is any curve from p to q. Then

$$\varphi(q) - \varphi(p) = \int_{\gamma[p,q]} \nabla \varphi(r) \cdot dr.$$

It is a generalization of the fundamental theorem of calculus to any curve in a plane or space (generally n-dimensional) rather than just the real line.

The gradient theorem implies that line integrals through gradient fields are path independent. In physics this theorem is one of the ways of defining a *conservative* force. By placing φ as potential, $\nabla \varphi$ is a conservative field. Work done by conservative forces does not depend on the path followed by the object, but only the end points, as the above equation shows.

The gradient theorem also has an interesting converse: any path-independent vector field can be expressed as the gradient of a scalar field. Just like the gradient theorem itself, this converse has many striking consequences and applications in both pure and applied mathematics.

Proof:

If φ is a differentiable function from some open subset U (of $\mathbb{R}^n$) to $\mathbb{R}$, and if r is a differentiable function from some closed interval $[a, b]$ to U, then by the multivariate chain rule, the composite function $\varphi \circ r$ is differentiable on (a, b) and

$$\frac{d}{dt}(\varphi \circ r)(t) = \nabla \varphi(r(t)) \cdot r'(t)$$

for all t in (a, b). Here the $\cdot$ denotes the usual inner product.

Now suppose the domain U of φ contains the differentiable curve γ with endpoints p and q, (oriented in the direction from p to q). If r parametrizes γ for t in $[a, b]$, then the above shows that

$$\int_\gamma \nabla\varphi(\mathrm{u})\cdot d\mathrm{u} = \int_a^b \nabla\varphi(\mathrm{r}(t))\cdot \mathrm{r}'(t)dt$$

$$= \int_a^b \frac{d}{dt}\varphi(\mathrm{r}(t))dt = \varphi(\mathrm{r}(b)) - \varphi(\mathrm{r}(a)) = \varphi(\mathrm{q}) - \varphi(\mathrm{p}),$$

where the definition of the line integral is used in the first equality, and the fundamental theorem of calculus is used in the third equality.

Example: Suppose $\gamma \subset \mathbb{R}^2$ is the circular arc oriented counterclockwise from $(5, 0)$ to $(-4, 3)$. Using the definition of a line integral,

$$\int_\gamma y dx + x dy = \int_0^{\pi - \tan^{-1}\left(\frac{3}{4}\right)} ((5\sin t)(-5\sin t) + (5\cos t)(5\cos t))dt$$

$$= \int_0^{\pi - \tan^{-1}\left(\frac{3}{4}\right)} 25\left(-\sin^2 t + \cos^2 t\right)dt$$

$$= \int_0^{\pi - \tan^{-1}\left(\frac{3}{4}\right)} 25\cos(2t)dt$$

$$= \frac{25}{2}\sin(2t)\Big|_0^{\pi - \tan^{-1}\left(\frac{3}{4}\right)}$$

$$= \frac{25}{2}\sin\left(2\pi - 2\tan^{-1}\left(\frac{3}{4}\right)\right)$$

$$= -\frac{25}{2}\sin\left(2\tan^{-1}\left(\frac{3}{4}\right)\right)$$

$$= -\frac{25\left(\frac{3}{4}\right)}{\left(\frac{3}{4}\right)^2 + 1} = -12.$$

Notice all of the painstaking computations involved in directly calculating the integral. Instead, since the function $f(x, y) = xy$ is differentiable on all of $\mathbb{R}^2$, we can simply use the gradient theorem to say,

$$\int_\gamma y dx + x dy = \int_\gamma \nabla(xy)\cdot(dx, dy) = xy\Big|_{(5,0)}^{(-4,3)} = -4\cdot 3 - 5\cdot 0 = -12.$$

Notice that either way gives the same answer, but using the latter method, most of the work is already done in the proof of the gradient theorem.

Example: For a more abstract example, suppose $\gamma \subset \mathbb{R}^n$ has endpoints p, q, with orientation from p to q. For u in $\mathbb{R}^n$, let $|u|$ denote the Euclidean norm of u. If $\alpha \geq 1$ is a real number, then

$$\int_\gamma \mathbf{x} \, |\mathbf{x}|^{\alpha-1} \, \mathbf{x} \cdot d\mathbf{x} = \frac{1}{\alpha+1} \int_\gamma (\alpha+1) \, | \, \mathbf{x} \, |^{(\alpha+1)-2} \, \mathbf{x} \cdot d\mathbf{x}$$

$$= \frac{1}{\alpha+1} \int_\gamma \nabla | \, \mathbf{x} \, |^{\alpha+1} \cdot d\mathbf{x} = \frac{| \, \mathbf{q} \, |^{\alpha+1} - | \, \mathbf{p} \, |^{\alpha+1}}{\alpha+1}$$

Here the final equality follows by the gradient theorem, since the function $f(\mathbf{x}) = |\mathbf{x}|^{\alpha+1}$ is differentiable on $\mathbb{R}^n$ if $\alpha \geq 1$.

If $\alpha < 1$ then this equality will still hold in most cases, but caution must be taken if γ passes through or encloses the origin, because the integrand vector field $|\mathbf{x}|^{\alpha-1}\mathbf{x}$ will fail to be defined there. However, the case $\alpha = -1$ is somewhat different; in this case, the integrand becomes $|\mathbf{x}|^{-2}\mathbf{x} = \nabla(\log |\mathbf{x}|)$, so that the final equality becomes $\log |\mathbf{q}| - \log |\mathbf{p}|$.

Note that if $n = 1$, then this example is simply a slight variant of the familiar power rule from single-variable calculus.

Example: Suppose there are n point charges arranged in three-dimensional space, and the i-th point charge has charge Q_i and is located at position p_i in $\mathbb{R}^3$. We would like to calculate the work done on a particle of charge q as it travels from a point a to a point b in $\mathbb{R}^3$. Using Coulomb's law, we can easily determine that the force on the particle at position r will be,

$$\mathbf{F}(\mathbf{r}) = kq \sum_{i=1}^n \frac{Q_i(\mathbf{r}-\mathbf{p}_i)}{|\mathbf{r}-\mathbf{p}_i|^3}$$

Here $|u|$ denotes the Euclidean norm of the vector u in $\mathbb{R}^3$, and $k = 1/(4\pi\varepsilon_0)$, where ε_0 is the vacuum permittivity.

Let $\gamma \subset \mathbb{R}^3 - \{p_1, ..., p_n\}$ be an arbitrary differentiable curve from a to b. Then the work done on the particle is,

$$W = \int_\gamma \mathbf{F}(\mathbf{r}) \cdot d\mathbf{r} = \int_\gamma \left(kq \sum_{i=1}^n \frac{Q_i(\mathbf{r}-\mathbf{p}_i)}{|\mathbf{r}-\mathbf{p}_i|^3} \right) \cdot d\mathbf{r} = kq \sum_{i=1}^n \left(Q_i \int_\gamma \frac{\mathbf{r}-\mathbf{p}_i}{|\mathbf{r}-\mathbf{p}_i|^3} \cdot d\mathbf{r} \right)$$

Now for each i, direct computation shows that,

$$\frac{\mathbf{r}-\mathbf{p}_i}{|\mathbf{r}-\mathbf{p}_i|^3} = -\nabla \frac{1}{|\mathbf{r}-\mathbf{p}_i|}.$$

Thus, continuing from above and using the gradient theorem,

$$W = -kq \sum_{i=1}^n \left(Q_i \int_\gamma \nabla \frac{1}{|\mathbf{r}-\mathbf{p}_i|} \cdot d\mathbf{r} \right) = kq \sum_{i=1}^n Q_i \left(\frac{1}{|\mathbf{a}-\mathbf{p}_i|} - \frac{1}{|\mathbf{b}-\mathbf{p}_i|} \right)$$

We are finished. Of course, we could have easily completed this calculation using the powerful language of electrostatic potential or electrostatic potential energy (with the familiar formulas

$W = -\Delta U = -q\Delta V$). However, we have not yet *defined* potential or potential energy, because the *converse* of the gradient theorem is required to prove that these are well-defined, differentiable functions and that these formulas hold. Thus, we have solved this problem using only Coulomb's Law, the definition of work, and the gradient theorem.

Converse of the Gradient Theorem

The gradient theorem states that if the vector field F is the gradient of some scalar-valued function (i.e., if F is conservative), then F is a path-independent vector field (i.e., the integral of F over some piecewise-differentiable curve is dependent only on end points). This theorem has a powerful converse:

If F is a path-independent vector field, then F is the gradient of some scalar-valued function.

It is straightforward to show that a vector field is path-independent if and only if the integral of the vector field over every closed loop in its domain is zero. Thus the converse can alternatively be stated as follows: If the integral of F over every closed loop in the domain of F is zero, then F is the gradient of some scalar-valued function.

Proof of the Converse

Suppose U is an open, path-connected subset of $\mathbb{R}^n$, and $F : U \to \mathbb{R}^n$ is a continuous and path-independent vector field. Fix some element a of U, and define $f : U \to \mathbb{R}$ by,

$$f(x) := \int_{\gamma[a,x]} F(u) \cdot du$$

Here $\gamma[a, x]$ is any (differentiable) curve in U originating at a and terminating at x. We know that f is well-defined because F is path-independent.

Let v be any nonzero vector in $\mathbb{R}^n$. By the definition of the directional derivative,

$$\frac{\partial f}{\partial v}(x) = \lim_{t \to 0} \frac{f(x+tv) - f(x)}{t}$$

$$= \lim_{t \to 0} \frac{\int_{\gamma[a,x+tv]} F(u) \cdot du - \int_{\gamma[a,x]} F(u) \cdot du}{t}$$

$$= \lim_{t \to 0} \frac{1}{t} \int_{\gamma[x,x+tv]} F(u) \cdot du$$

To calculate the integral within the final limit, we must parametrize $\gamma[x, x + tv]$. Since F is path-independent, U is open, and t is approaching zero, we may assume that this path is a straight line, and parametrize it as u(s) = x + sv for $0 < s < t$. Now, since u'(s) = v, the limit becomes:

$$\lim_{t \to 0} \frac{1}{t} \int_0^t F(u(s)) \cdot u'(s) ds = \frac{d}{dt} \int_0^t F(x + s\, v) \cdot v\, ds \Big|_{t=0} = F(x) \cdot v$$

Thus we have a formula for $\partial_v f$, where v is arbitrary. Let x = $(x_1, x_2, ..., x_n)$ and let e_i denote the i-th standard basis vector so that,

$$\nabla f(\mathrm{x}) = \left(\frac{\partial f(\mathrm{x})}{\partial x_1}, \frac{\partial f(\mathrm{x})}{\partial x_2}, ..., \frac{\partial f(\mathrm{x})}{\partial x_n}\right) = (\mathrm{F}(\mathrm{x}) \cdot \mathrm{e}_1, \mathrm{F}(\mathrm{x}) \cdot \mathrm{e}_2, ..., \mathrm{F}(\mathrm{x}) \cdot \mathrm{e}_n) = \mathrm{F}(\mathrm{x})$$

Thus we have found a scalar-valued function f whose gradient is the path-independent vector field F, as desired.

Example of the Converse Principle

To illustrate the power of this converse principle, we cite an example that has significant physical consequences. In classical electromagnetism, the electric force is a path-independent force; i.e. the work done on a particle that has returned to its original position within an electric field is zero (assuming that no changing magnetic fields are present).

Therefore, the above theorem implies that the electric force field $F_e : S \to \mathbb{R}^3$ is conservative (here S is some open, path-connected subset of $\mathbb{R}^3$ that contains a charge distribution). Following the ideas of the above proof, we can set some reference point a in S, and define a function $U_e : S \to \mathbb{R}$ by

$$U_e(\mathrm{r}) := -\int_{\gamma[a,r]} \mathrm{F}_e(\mathrm{u}) \cdot \mathrm{du}$$

Using the above proof, we know U_e is well-defined and differentiable, and $F_e = -\nabla U_e$ (from this formula we can use the gradient theorem to easily derive the well-known formula for calculating work done by conservative forces: $W = -\Delta U$). This function U_e is often referred to as the electrostatic potential energy of the system of charges in S (with reference to the zero-of-potential a). In many cases, the domain S is assumed to be unbounded and the reference point a is taken to be "infinity", which can be made rigorous using limiting techniques. This function U_e is an indispensable tool used in the analysis of many physical systems.

Generalizations

Many of the critical theorems of vector calculus generalize elegantly to statements about the integration of differential forms on manifolds. In the language of differential forms and exterior derivatives, the gradient theorem states that

$$\int_{\partial\gamma} \phi = \int_{\gamma} d\phi$$

for any 0-form, ϕ, defined on some differentiable curve $\gamma \subset \mathbb{R}^n$ (here the integral of ϕ over the boundary of the γ is understood to be the evaluation of ϕ at the endpoints of γ).

Notice the striking similarity between this statement and the generalized version of Stokes' theorem, which says that the integral of any compactly supported differential form ω over the boundary of some orientable manifold Ω is equal to the integral of its exterior derivative $d\omega$ over the whole of Ω, i.e.,

$$\int_{\partial\Omega} \omega = \int_{\Omega} d\omega$$

This powerful statement is a generalization of the gradient theorem from 1-forms defined on one-dimensional manifolds to differential forms defined on manifolds of arbitrary dimension.

The converse statement of the gradient theorem also has a powerful generalization in terms of differential forms on manifolds. In particular, suppose ω is a form defined on a contractible domain, and the integral of ω over any closed manifold is zero. Then there exists a form ψ such that $\omega = d\psi$. Thus, on a contractible domain, every closed form is exact. This result is summarized by the Poincaré lemma.

SQUEEZE THEOREM

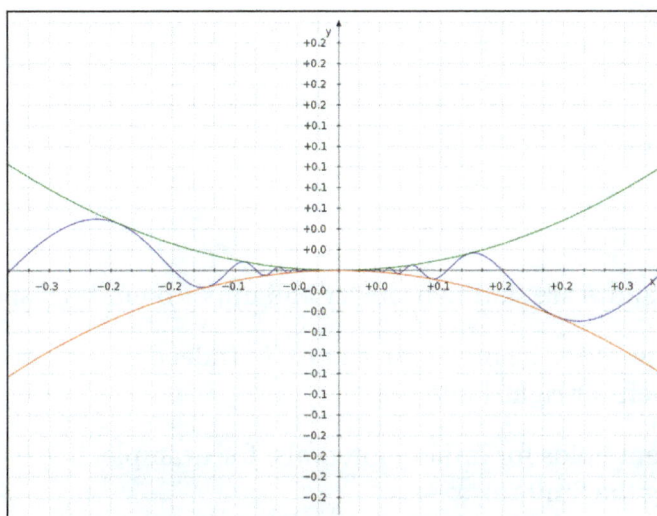

Illustration of the squeeze theorem

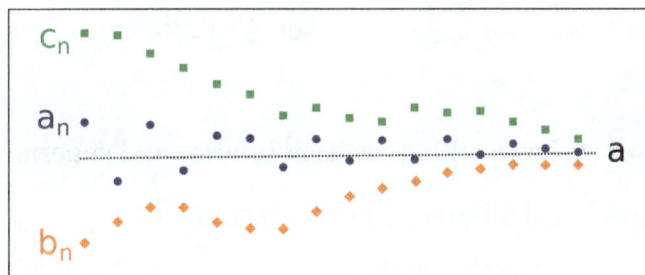

When a sequence lies between two other converging sequences with the same limit, it also converges to this limit.

In calculus, the squeeze theorem, also known as the pinching theorem, the sandwich theorem, the sandwich rule, and sometimes the squeeze lemma, is a theorem regarding the limit of a function. In Italy, the theorem is also known as theorem of Carabinieri.

The squeeze theorem is used in calculus and mathematical analysis. It is typically used to confirm the limit of a function via comparison with two other functions whose limits are known or easily computed. It was first used geometrically by the mathematicians Archimedes and Eudoxus in an effort to compute π, and was formulated in modern terms by Carl Friedrich Gauss.

In many languages (e.g. French, German, Italian and Russian), the squeeze theorem is also known as the two policemen (and a drunk) theorem, or some variation thereof. The story is that if two policemen are escorting a drunk prisoner between them, and both officers go to a cell, then (regardless of the path taken, and the fact that the prisoner may be wobbling about between the policemen) the prisoner must also end up in the cell.

Statement

The squeeze theorem is formally stated as follows.

Let I be an interval having the point a as a limit point. Let $g, f,$ and h be functions defined on I, except possibly at a itself. Suppose that for every x in I not equal to a we have,

$$g(x) \leq f(x) \leq h(x)$$

and also suppose that,

$$\lim_{x \to a} g(x) = \lim_{x \to a} h(x) = L.$$

Then $\lim_{x \to a} f(x) = L.$

- The functions g and h are said to be lower and upper bounds (respectively) of f.

- Here, a is *not* required to lie in the interior of I. Indeed, if a is an endpoint of I, then the above limits are left- or right-hand limits.

- A similar statement holds for infinite intervals: for example, if $I = (0, \infty)$, then the conclusion holds, taking the limits as $x \to \infty$.

This theorem is also valid for sequences. Let $(a_n), (c_n)$ be two sequences converging to ℓ, and (b_n) a sequence. If $\forall n \geqslant N, N \in \mathbb{N}$ we have $a_n \leqslant b_n \leqslant c_n$, then (b_n) also converges to ℓ.

Proof:

From the above hypotheses we have, taking the limit inferior and superior:

$$L = \lim_{x \to a} g(x) \leq \liminf_{x \to a} f(x) \leq \limsup_{x \to a} f(x) \leq \lim_{x \to a} h(x) = L,$$

so all the inequalities are indeed equalities, and the thesis immediately follows.

A direct proof, using the (ϵ, δ)-definition of limit, would be to prove that for all real $\epsilon > 0$ there exists a real $\delta > 0$ such that for all x with $| x - a | < \delta$, we have $| f(x) - L | < \epsilon$. Symbolically,

$$\forall \epsilon > 0, \exists \delta > 0 : \forall x, (| x - a | < \delta \Rightarrow | f(x) - L | < \epsilon)$$

As,

$$\lim_{x \to a} g(x) = L$$

means that,

$$\forall \varepsilon > 0, \exists \delta_1 > 0 : \forall x \, (|x-a| < \delta_1 \Rightarrow |g(x)-L| < \varepsilon).$$

and

$$\lim_{x \to a} h(x) = L$$

means that,

$$\forall \varepsilon > 0, \exists \delta_2 > 0 : \forall x \, (|x-a| < \delta_2 \Rightarrow |h(x)-L| < \varepsilon),$$

then we have,

$$g(x) \leq f(x) \leq h(x)$$

$$g(x)-L \leq f(x)-L \leq h(x)-L$$

We can choose $\delta := \min\{\delta_1, \delta_2\}$. Then, if $|x-a| < \delta$, combining $\forall \varepsilon > 0, \exists \delta_1 > 0 : \forall x \, (|x-a| < \delta_1 \Rightarrow |g(x)-L| < \varepsilon)$ and $\forall \varepsilon > 0, \exists \delta_2 > 0 : \forall x \, (|x-a| < \delta_2 \Rightarrow |h(x)-L| < \varepsilon)$, we have

$$-\varepsilon < g(x)-L \leq f(x)-L \leq h(x)-L < \varepsilon,$$

$$-\varepsilon < f(x)-L < \varepsilon,$$

which completes the proof.

The proof for sequences is very similar, using the $\in$-definition of a limit of a sequence.

Statement for Series

There is also the squeeze theorem for series, which can be stated as follows:

Let $\sum_n a_n, \sum_n c_n$ be two convergent series. If $\exists N \in \mathbb{N}$ such that $\forall n > N, a_n \leqslant b_n \leqslant c_n$ then $\sum_n b_n$ also converges.

Proof:

Let $\sum_n a_n, \sum_n c_n$ be two convergent series. Hence, the sequences $\left(\sum_{k=1}^{n} a_n\right)_{n=1}^{\infty}, \left(\sum_{k=1}^{n} c_n\right)_{n=1}^{\infty}$ are Cauchy. That is, for fixed $\epsilon > 0$,

$$\exists N_1 \in \mathbb{N} \text{ such that } \forall n > m > N_1, \left|\sum_{k=1}^{n} a_n - \sum_{k=1}^{m} a_n\right| < \epsilon \Rightarrow \left|\sum_{k=m+1}^{n} a_n\right| < \epsilon \Rightarrow -\epsilon < \sum_{k=m+1}^{n} a_n < \epsilon$$

and similarly $\exists N_2 \in \mathbb{N}$ such that $\forall n > m > N_2, \left|\sum_{k=1}^{n} c_n - \sum_{k=1}^{m} c_n\right| < \epsilon \Rightarrow \left|\sum_{k=m+1}^{n} c_n\right| < \epsilon \Rightarrow -\epsilon < \sum_{k=m+1}^{n} c_n < \epsilon.$

We know that $\exists N_3 \in \mathbb{N}$ such that $\forall n > N_1, a_n \leqslant b_n \leqslant c_n$. Hence, $\forall n > m > \max\{N_1, N_2, N_3\}$, we have combining the above two equations:

$$a_n \leqslant b_n \leqslant c_n \Rightarrow \sum_{k=m+1}^{n} a_n \leqslant \sum_{k=m+1}^{n} b_n \leqslant \sum_{k=m+1}^{n} c_n \Rightarrow -\epsilon < \sum_{k=m+1}^{n} b_n < \epsilon \Rightarrow \left| \sum_{k=m+1}^{n} b_n \right| < \epsilon \Rightarrow \left| \sum_{k=1}^{n} b_n - \sum_{k=1}^{m} b_n \right| < \epsilon.$$

Therefore $\left(\sum_{k=1}^{n} b_n \right)_{n=1}^{\infty}$ is a Cauchy sequence. So $\sum_{n} b_n$ converges.

Example:

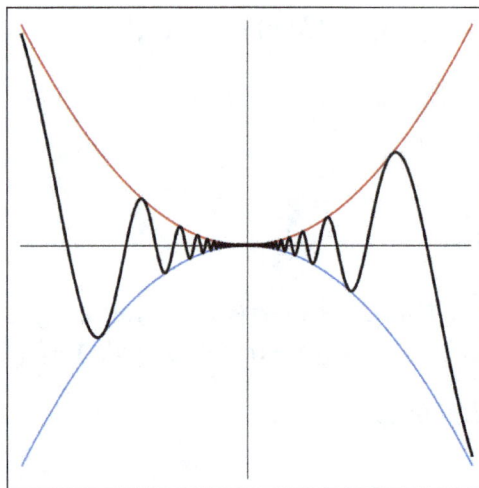

$x^2 \sin(1/x)$ being squeezed in the limit as x goes to 0

The limit,

$$\lim_{x \to 0} x^2 \sin(\tfrac{1}{x})$$

cannot be determined through the limit law,

$$\lim_{x \to a} (f(x) \cdot g(x)) = \lim_{x \to a} f(x) \cdot \lim_{x \to a} g(x),$$

because,

$$\lim_{x \to 0} \sin(\tfrac{1}{x})$$

does not exist.

However, by the definition of the sine function,

$$-1 \leq \sin(\tfrac{1}{x}) \leq 1.$$

It follows that,

$$-x^2 \leq x^2 \sin(\tfrac{1}{x}) \leq x^2$$

Since $\lim\limits_{x \to 0} -x^2 = \lim\limits_{x \to 0} x^2 = 0$, by the squeeze theorem, $\lim\limits_{x \to 0} x^2 \sin(\tfrac{1}{x})$ must also be 0.

Example:

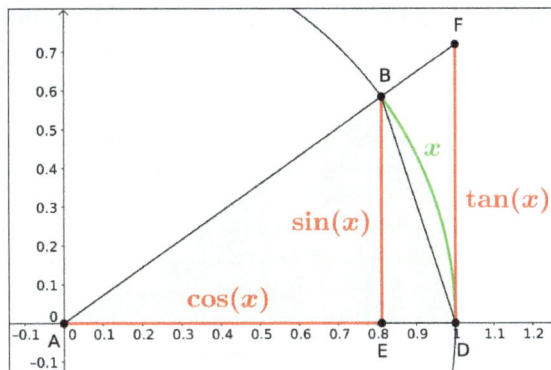

Comparing areas:

$$A(\triangle ADF) \geq A(\text{sector } ADB) \geq A(\triangle ADB)$$

$$\Rightarrow \frac{1}{2} \cdot \tan(x) \cdot 1 \geq \frac{x}{2\pi} \cdot \pi \geq \frac{1}{2} \cdot \sin(x) \cdot 1$$

$$\Rightarrow \frac{\sin(x)}{\cos(x)} \geq x \geq \sin(x)$$

$$\Rightarrow \frac{\cos(x)}{\sin(x)} \leq \frac{1}{x} \leq \frac{1}{\sin(x)}$$

$$\Rightarrow \cos(x) \leq \frac{\sin(x)}{x} \leq 1$$

Probably the best-known examples of finding a limit by squeezing are the proofs of the equalities,

$$\lim_{x \to 0} \frac{\sin(x)}{x} = 1,$$

$$\lim_{x \to 0} \frac{1 - \cos(x)}{x} = 0.$$

The first limit follows by means of the squeeze theorem from the fact that,

$$\cos x \leq \frac{\sin(x)}{x} \leq 1$$

for x close enough to 0. The correctness of which for positive x can be seen by simple geometric reasoning that can be extended to negative x as well. The second limit follows from the squeeze theorem and the fact that,

$$0 \leq \frac{1 - \cos(x)}{x} \leq x$$

for x close enough to 0. This can be derived by replacing $\sin(x)$ in the earlier fact by $\sqrt{1-\cos(x)^2}$ and squaring the resulting inequality.

These two limits are used in proofs of the fact that the derivative of the sine function is the cosine function. That fact is relied on in other proofs of derivatives of trigonometric functions.

Example:

It is possible to show that,

$$\frac{d}{d\theta}\tan\theta = \sec^2\theta$$

by squeezing, as follows.

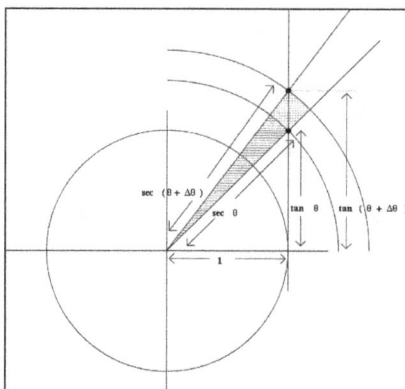

In the illustration at right, the area of the smaller of the two shaded sectors of the circle is,

$$\frac{\sec^2\theta\Delta\theta}{2},$$

since the radius is $\sec\theta$ and the arc on the unit circle has length $\Delta\theta$. Similarly, the area of the larger of the two shaded sectors is,

$$\frac{\sec^2(\theta+\Delta\theta)\Delta\theta}{2}.$$

What is squeezed between them is the triangle whose base is the vertical segment whose endpoints are the two dots. The length of the base of the triangle is $\tan(\theta + \Delta\theta) - \tan(\theta)$, and the height is 1. The area of the triangle is therefore,

$$\frac{\tan(\theta+\Delta\theta)-\tan(\theta)}{2}.$$

From the inequalities,

$$\frac{\sec^2\theta\Delta\theta}{2} \leq \frac{\tan(\theta+\Delta\theta)-\tan(\theta)}{2} \leq \frac{\sec^2(\theta+\Delta\theta)\Delta\theta}{2}$$

we deduce that,

$$\sec^2\theta \le \frac{\tan(\theta + \Delta\theta) - \tan(\theta)}{\Delta\theta} \le \sec^2(\theta + \Delta\theta),$$

provided $\Delta\theta > 0$, and the inequalities are reversed if $\Delta\theta < 0$. Since the first and third expressions approach $\sec^2\theta$ as $\Delta\theta \to 0$, and the middle expression approaches $(d/d\theta)\tan\theta$, the desired result follows.

Example: The squeeze theorem can still be used in multivariable calculus but the lower (and upper functions) must be below (and above) the target function not just along a path but around the entire neighborhood of the point of interest and it only works if the function really does have a limit there. It can, therefore, be used to prove that a function has a limit at a point, but it can never be used to prove that a function does not have a limit at a point.

$$\lim_{(x,y) \to (0,0)} \frac{x^2 y}{x^2 + y^2}$$

cannot be found by taking any number of limits along paths that pass through the point, but since

$$0 \le \frac{x^2}{x^2 + y^2} \le 1$$

$$-|y| \le y \le |y|$$

$$-|y| \le \frac{x^2 y}{x^2 + y^2} \le |y|$$

$$\lim_{(x,y) \to (0,0)} -|y| = 0$$

$$\lim_{(x,y) \to (0,0)} |y| = 0$$

$$0 \le \lim_{(x,y) \to (0,0)} \frac{x^2 y}{x^2 + y^2} \le 0$$

therefore, by the squeeze theorem,

$$\lim_{(x,y) \to (0,0)} \frac{x^2 y}{x^2 + y^2} = 0$$

INVERSE FUNCTION THEOREM

The Inverse Function Theorem. Let $f : \mathbb{R}^n \to \mathbb{R}^n$ be continuously differentiable on some open set containing a, and suppose $\det Jf(a) \ne 0$. Then there is some open set V containing a and an open

W containing f(a) such that $f : V \to W$ has a continuous inverse $f^{-1} : W \to V$ which is differentiable for all y ∈ W.

As matrices, $J\left(f^{-1}\right)(y) = \left[\left(Jf\right)\left(f^{-1}\left(y\right)\right)\right]^{-1}$.

Lemma: Let $A \subset \mathbb{R}^n$ be an open rectangle, and suppose $f : A \to \mathbb{R}^n$ is continuously differentiable. If there is some M > 0 such that,

$$\left|\frac{\partial f_i}{\partial x_j}(x)\right| \le M, \forall x \in A, \text{then } \left\|f(y) - f(z)\right\| \le n^2 \cdot M \cdot \left\|y - z\right\|, \forall y, z \in A.$$

Proof: We write,

$$f_i(y) - f_i(z) = f_i(y_1, \ldots, y_n) - f_i(z_1, \ldots, z_n)$$

$$= \sum_{j=1}^{n}\left[f\left(y_1, \ldots, y_j, z_{j+1}, \ldots, z_n\right) - f\left(y_1, \ldots, y_{j-1}, z_j, z_{j+1}, \ldots, z_n\right)\right]$$

$$= \sum_{j=1}^{n}\frac{\partial f_i}{\partial x_j}\left(\mathrm{x}_{ij}\right)\left(y_j - z_j\right)$$

for some $\mathrm{x}_{ij} = \left(y_1, \ldots, y_{j-1}, c_j, z_{j+1}, \ldots, z_n\right)$ where, for each j = 1, . . . , n, we have c_j is in the interval (y_j, z_j), by the single-variable Mean Value Theorem.

Then,

$$\left\|f(y) - f(z)\right\| \le \sum_{i=1}^{n}\left\|f_i(y) - f_i(z)\right\|$$

$$= \sum_{i=1}^{n}\sum_{j=1}^{n}\left|\frac{\partial f_i}{\partial x_j}\left(\mathrm{x}_{ij}\right)\right| \cdot \left|y_j - z_j\right|$$

$$\le \sum_{i=1}^{n}\sum_{j=1}^{n}M \cdot \left\|y - z\right\|$$

$$= n^2 \cdot M \cdot \left\|y - z\right\|$$

Proof of the Inverse Function Theorem

Let $L = Jf(a)$. Then $\det(L) \ne 0$, and so L^{-1} exists. Consider the composite function L^{-1} of $f : \mathbb{R}^n \to \mathbb{R}^n$ Then:

$$\left(L^{-1} \circ f\right)(a) = J\left(L^{-1}\right)\left(f(a)\right) \circ Jf(a)$$

$$= L^{-1} \circ J f(a)$$

$$= L^{-1} \circ L$$

which is the identity. Since L is invertible, the theorem is equally true or false for both $L^{-1} \circ f$ and f simultaneously, and hence we prove it in the case when L = I.

Suppose $f(a + h) = f(a)$. Then $\dfrac{|f(a+h) - f(a) - L(h)|}{|h|} = \dfrac{|h|}{|h|} = 1$

On the other hand, we have have $\lim\limits_{\|h\| \to 0} \dfrac{f(a+h) - f(a) - L(h)}{\|h\|} = 0,$

which is a contradiction, and hence there must be some open neighborhood/rectangle U around a in which $f(a + h) \neq f(a)$, $\forall a + h \in U$, $h \neq 0$.

Furthermore, we may choose this neighborhood U small enough so that:

$$\det\left(Jf(x) \right) \neq 0, \ \forall x \in U$$

$$\left| \frac{\partial f_i}{\partial x_j}(x) - \frac{\partial f_i}{\partial x_j}(a) \right| < \frac{1}{2n^2}, \forall i, j, \ \forall x \in U$$

since these are conditions on $n^2 + 1$ continuous functions.

Claim: $\|x_1 - x_2\| \leq 2 \cdot \|f(x_1) - f(x_2)\|$, $\forall x_1, x_2 \in U$

Proof of Claim: First, we let g (x) = f (x) − x. By construction and

the second fact above, we have $\left| \dfrac{\partial g_i}{\partial x_j}(x) \right| = \left| \dfrac{\partial f_i}{\partial x_j}(x) - \dfrac{\partial f_i}{\partial x_j}(a) \right| \leq \dfrac{1}{2n^2},$

and so we apply the Lemma with $M = \dfrac{1}{2n^2}$;

$$\|x_1 - x_2\| - \|f(x_1) - f(x_2)\| \ \leq \ \|(f(x_1) - x_1) - (f(x_2) - x_2)\|$$
$$= \|g(x_1) - g(x_2)\|$$
$$\leq \frac{1}{2} \cdot \|x_1 - x_2\|$$

and so, combining these inequalities we have,

$$\frac{1}{2} \cdot \|x_1 - x_2\| \leq \|f(x_1) - f(x_2)\|$$

Now consider the set ∂U, which is compact since U is bounded. We know by the reasoning in the second paragraph of the proof that if $x \in \partial U$ then $f(x) \neq f(a)$. Hence $\exists d > 0$ such that $\|f(x) - f(a)\| \geq d$, $\forall x \in \partial U$. (Since both f and the taking of norms are continuous functions, the expression $\|f(x) - f(a)\|$ attains its non-zero minimum on the compact set ∂U.)

We construct the set $W \subset \mathbb{R}^n$ thinking of it as a subset of the range of f, as follows:

$$W = \left\{ y \in \mathbb{R}^n \Big| \|y - f(a)\| < \frac{d}{2} \right\} = B_{d/2}(f(a))$$

By its construction and the use of the positive real number d, we see that if $y \in W$ and $x \in \partial U$, then

$$\|y - f(a)\| < \|y - f(x)\|.$$

Claim: Given $y \in W$, there is a unique $x \in U$ such that f(x) = y.

Proof of Claim:

Existence:

Consider $h : U \to \mathbb{R}$ defined by $h(x) = \|y - f(x)\|^2$. A straightforward simplification of this expression gives $h(x) = \sum_{i=1}^{n} (y_i - f_i(x))^2$.

That h is continuous and hence attains its minimum on the compact set $\overline{U}$. This minimum does not occur on the boundary, ∂U, by the inequality and hence it must occur on the interior. Since h is also differentiable, we must have $\nabla h(x) = 0$ at the minimum, and hence:

$$0 = \frac{\partial h}{\partial x_j}(x) = \sum_{i=1}^{n} 2 \cdot (y_i - f_i(x)) \cdot \frac{\partial f_i}{\partial x_j}(x), \ \forall j$$

In other words, collecting this information over the various i and j we have,

$$0 = Jf(x) \cdot (y - f(x)),$$

but since we have assumed that $\det Jf(x) \neq 0$ for any $x \in U$, it follows that Jf(x) is invertible, and hence y − f(x) = 0.

Uniqueness:

We use Claim 1. Suppose $y = f(x_1) = f(x_2)$.

Then $\|x_1 - x_2\| \leq 2 \cdot \|f(x_1) - f(x_2)\| = 0$, and hence $x_1 = x_2$.

By Claim 2, if we define V = U ∩ f −1 (W), then f : V → U has an inverse. It remains to show that f⁻¹ is continuous and differentiable. Even though continuity would follow from differentiability, we do this in two steps because we will use the continuity to help prove the differentiability.

Claim: f^{-1} is continuous.

Proof of Claim:

For $y_1, y_2 \in W$, find $x_1, x_2 \in U$ such that $f(x_1) = y_1$ and $f(x_2) = y_2$. Claim 1 implies that $\|x_1 - x_2\| \leq 2 \cdot \|f(x_1) - f(x_2)\|$, or in other words, that $\|f^{-1}(y_1) - f^{-1}(y_2)\| \leq 2 \cdot \|y_1 - y_2\|$.

It is now easy to see that given ε > 0, we need only choose δ = ε/2 to guarantee that if $||y_1 - y_2|| <$ δ, then $||f^{-1}(y_1) - f^{-1}(y_2)|| < ε$.

Claim: f^{-1} is differentiable.

Proof of Claim:

Let x ∈ V, let A = J f (x), and let y = f (x) ∈ W.

We claim that $J f^{-1}(y) = A^{-1}$.

Define φ (x) = f (x + h) − f (x) − A (h).

Then $\lim\limits_{||h|| \to 0} \dfrac{||\varphi(h)||}{||h||} = 0$, by the differentiability of f.

Since det (A) = det $J f(x) \neq 0$ by hypothesis, we know that A^{-1} exists, and it is linear since A is. Then:

$$A^{-1}\left(f(x+h) - f(x)\right) = h + A^{-1}\left(\varphi(h)\right)$$
$$= \left[(x+h) - x\right] + A^{-1}\left(\varphi(h)\right)$$

Letting y = f (x) and y_1 = f (x + h) on both sides yields:

$$A^{-1}(y_1 - y) = \left[f^{-1}(y_1) - f^{-1}(y)\right] + A^{-1}\left(\varphi(f^{-1}(y_1) - f^{-1}(y))\right)$$

Re-arranging sides:

$$A^{-1}\left(\varphi(f^{-1}(y_1) - f^{-1}(y))\right) = \left[f^{-1}(y_1) - f^{-1}(y)\right] - A^{-1}(y_1 - y)$$

To show differentiability, we need:

$$\lim_{||y_1 - y|| \to 0} \frac{\left\|f^{-1}(y_1) - f^{-1}(y) - A^{-1}(y_1 - y)\right\|}{||y_1 - y||} = 0$$

but by equation $A^{-1}\left(\varphi(f^{-1}(y_1) - f^{-1}(y))\right) = \left[f^{-1}(y_1) - f^{-1}(y)\right] - A^{-1}(y_1 - y)$ above, this is the same as showing:

$$\lim_{||y_1 - y|| \to 0} \frac{\left\|A^{-1}\left(\varphi f^{-1}(y_1) - f^{-1}(y)\right)\right\|}{||y_1 - y||} = 0.$$

Since A^{-1} is linear, it suffices to use the Chain Rule and show that:

$$\lim_{||y_1 - y|| \to 0} \frac{\left\|\varphi f^{-1}(y_1) - f^{-1}(y)\right\|}{||y_1 - y||} = 0.$$

so we factor the expression inside the limit as follows:

$$\frac{\left\|\varphi f^{-1}\left(y_{1}\right)-f^{-1}\left(y\right)\right\|}{\left\|y_{1}-y\right\|}=\frac{\left\|\varphi f^{-1}\left(y_{1}\right)-f^{-1}\left(y\right)\right\|}{\left\|f^{-1}\left(y_{1}\right)-f^{-1}\left(y\right)\right\|}\cdot\frac{\left\|f^{-1}\left(y_{1}\right)-f^{-1}\left(y\right)\right\|}{\left\|y_{1}-y\right\|}.$$

The first term on the right tends to 0 because of how we defined φ and the fact that the continuity of f^{-1} means that $f^{-1}(y_1) \to f^{-1}(y)$.

Observing that the second term on the right is less than or equal to 2 (by Claim 1) enables us to use the Squeeze Theorem and conclude that the product on the right tends to 0, which establishes equation

$$\lim_{\|y_{1}-y\|\to 0}\frac{\left\|\varphi f^{-1}\left(y_{1}\right)-f^{-1}\left(y\right)\right\|}{\left\|y_{1}-y\right\|}=0.$$

References

- Bartle, Robert G.; Sherbert, Donald R. (2011), Introduction to Real Analysis (4th ed.), Wiley, ISBN 978-0-471-43331-6

- Ballantine, C.; Roberts, J. (January 2002), "A Simple Proof of Rolle's Theorem for Finite Fields", The American Mathematical Monthly, Mathematical Association of America, 109 (1): 72–74, doi:10.2307/2695770, JSTOR 2695770

- Fundamental-theorem-of-calculus, integrals, maths, guides: toppr.com, Retrieved 31 August, 2019

- Keisler, H. Jerome (1986). Elementary Calculus : An Infinitesimal Approach (PDF). Boston: Prindle, Weber & Schmidt. p. 164. ISBN 0-87150-911-3

- Jittorntrum, K. (1978). "An Implicit Function Theorem". Journal of Optimization Theory and Applications. 25 (4): 575–577. doi:10.1007/BF00933522

PERMISSIONS

All chapters in this book are published with permission under the Creative Commons Attribution Share Alike License or equivalent. Every chapter published in this book has been scrutinized by our experts. Their significance has been extensively debated. The topics covered herein carry significant information for a comprehensive understanding. They may even be implemented as practical applications or may be referred to as a beginning point for further studies.

We would like to thank the editorial team for lending their expertise to make the book truly unique. They have played a crucial role in the development of this book. Without their invaluable contributions this book wouldn't have been possible. They have made vital efforts to compile up to date information on the varied aspects of this subject to make this book a valuable addition to the collection of many professionals and students.

This book was conceptualized with the vision of imparting up-to-date and integrated information in this field. To ensure the same, a matchless editorial board was set up. Every individual on the board went through rigorous rounds of assessment to prove their worth. After which they invested a large part of their time researching and compiling the most relevant data for our readers.

The editorial board has been involved in producing this book since its inception. They have spent rigorous hours researching and exploring the diverse topics which have resulted in the successful publishing of this book. They have passed on their knowledge of decades through this book. To expedite this challenging task, the publisher supported the team at every step. A small team of assistant editors was also appointed to further simplify the editing procedure and attain best results for the readers.

Apart from the editorial board, the designing team has also invested a significant amount of their time in understanding the subject and creating the most relevant covers. They scrutinized every image to scout for the most suitable representation of the subject and create an appropriate cover for the book.

The publishing team has been an ardent support to the editorial, designing and production team. Their endless efforts to recruit the best for this project, has resulted in the accomplishment of this book. They are a veteran in the field of academics and their pool of knowledge is as vast as their experience in printing. Their expertise and guidance has proved useful at every step. Their uncompromising quality standards have made this book an exceptional effort. Their encouragement from time to time has been an inspiration for everyone.

The publisher and the editorial board hope that this book will prove to be a valuable piece of knowledge for students, practitioners and scholars across the globe.

INDEX